The Mathematical Sciences in 2025

Committee on the Mathematical Sciences in 2025

Board on Mathematical Sciences and Their Applications

Division on Engineering and Physical Sciences

NATIONAL RESEARCH COUNCIL
OF THE NATIONAL ACADEMIES

THE NATIONAL ACADEMIES PRESS
Washington, D.C.
www.nap.edu

THE NATIONAL ACADEMIES PRESS   500 Fifth Street, NW   Washington, DC 20001

NOTICE: The project that is the subject of this report was approved by the Governing Board of the National Research Council, whose members are drawn from the councils of the National Academy of Sciences, the National Academy of Engineering, and the Institute of Medicine. The members of the committee responsible for the report were chosen for their special competences and with regard for appropriate balance.

This project was supported by the National Science Foundation under grant number DMS-0911899. Any opinions, findings, conclusions, or recommendations expressed in this publication are those of the author(s) and do not necessarily reflect the views of the organizations or agencies that provided support for the project.

International Standard Book Number-13: 978-0-309-28457-8
International Standard Book Number-10: 0-309-28457-0
Library of Congress Control Number: 2013933839

Additional copies of this report are available from the National Academies Press, 500 Fifth Street, NW, Keck 360, Washington, DC 20001; (800) 624-6242 or (202) 334-3313; http://www.nap.edu.

Suggested citation: National Research Council. 2013. *The Mathematical Sciences in 2025*. Washington, D.C.: The National Academies Press.

THE NATIONAL ACADEMIES
Advisers to the Nation on Science, Engineering, and Medicine

The **National Academy of Sciences** is a private, nonprofit, self-perpetuating society of distinguished scholars engaged in scientific and engineering research, dedicated to the furtherance of science and technology and to their use for the general welfare. Upon the authority of the charter granted to it by the Congress in 1863, the Academy has a mandate that requires it to advise the federal government on scientific and technical matters. Dr. Ralph J. Cicerone is president of the National Academy of Sciences.

The **National Academy of Engineering** was established in 1964, under the charter of the National Academy of Sciences, as a parallel organization of outstanding engineers. It is autonomous in its administration and in the selection of its members, sharing with the National Academy of Sciences the responsibility for advising the federal government. The National Academy of Engineering also sponsors engineering programs aimed at meeting national needs, encourages education and research, and recognizes the superior achievements of engineers. Dr. Charles M. Vest is president of the National Academy of Engineering.

The **Institute of Medicine** was established in 1970 by the National Academy of Sciences to secure the services of eminent members of appropriate professions in the examination of policy matters pertaining to the health of the public. The Institute acts under the responsibility given to the National Academy of Sciences by its congressional charter to be an adviser to the federal government and, upon its own initiative, to identify issues of medical care, research, and education. Dr. Harvey V. Fineberg is president of the Institute of Medicine.

The **National Research Council** was organized by the National Academy of Sciences in 1916 to associate the broad community of science and technology with the Academy's purposes of furthering knowledge and advising the federal government. Functioning in accordance with general policies determined by the Academy, the Council has become the principal operating agency of both the National Academy of Sciences and the National Academy of Engineering in providing services to the government, the public, and the scientific and engineering communities. The Council is administered jointly by both Academies and the Institute of Medicine. Dr. Ralph J. Cicerone and Dr. Charles M. Vest are chair and vice chair, respectively, of the National Research Council.

www.national-academies.org

COMMITTEE ON THE MATHEMATICAL SCIENCES IN 2025

THOMAS E. EVERHART, California Institute of Technology, *Chair*
MARK L. GREEN, University of California, Los Angeles, *Vice-chair*
TANYA S. BEDER, SBCC Group, Inc.
JAMES O. BERGER, Duke University
LUIS A. CAFFARELLI, University of Texas at Austin
EMMANUEL J. CANDES, Stanford University
PHILLIP COLELLA, E.O. Lawrence Berkeley National Laboratory
DAVID EISENBUD, University of California, Berkeley
PETER W. JONES, Yale University
JU-LEE KIM, Massachusetts Institute of Technology
YANN LeCUN, New York University
JUN LIU, Harvard University
JUAN MALDACENA, Institute for Advanced Study
JOHN W. MORGAN, Stony Brook University
YUVAL PERES, Microsoft Research
EVA TARDOS, Cornell University
MARGARET H. WRIGHT, New York University
JOE B. WYATT, Vanderbilt University

Staff

SCOTT WEIDMAN, Study Director
THOMAS ARRISON, Senior Program Officer
MICHELLE SCHWALBE, Program Officer
BARBARA WRIGHT, Administrative Assistant

Preface

When I was asked to chair a committee of mathematical scientists charged with examining the field now with an eye toward how it needs to evolve to produce the best value for the country by 2025, I demurred because I am not a mathematical scientist. The counter was that therefore I would not be biased, could be objective to prevent possible internal politics from "capturing" the report, and would be continuing a tradition of having such committees chaired by nonexperts. The assignment was educational in many ways.

The committee was extraordinary in its makeup, with experts from the core of mathematics as well as from departments of statistics and computer science, from both academia and industry. My eyes were opened to the power of the mathematical sciences today, not only as an intellectual undertaking in their own right but also as the increasingly modern foundation for much of science, engineering, medicine, economics, and business. The increasingly important challenges of deriving knowledge from huge amounts of data, whether numerical or experimental, of simulating complex phenomena accurately, and of dealing with uncertainty intelligently are some of the areas where mathematical scientists have important contributions to make going forward—and the members of this committee know it. They have demonstrated a great capacity to envision an emerging era in which the mathematical sciences underpin much of twenty-first century science, engineering, medicine, industry, and national security. I hope that this report persuades many others to embrace that vision.

While all members of the committee contributed to this report, vice-chair Mark Green, from the University of California at Los Angeles, and

NRC staff, headed by Scott Weidman, worked tirelessly to provide much of the writing and data that give the report its coherence, organization, and credibility. I especially thank them, for myself and for the rest of the committee, for their essential contributions.

Thomas E. Everhart, *Chair*

Acknowledgments

This report has been reviewed in draft form by individuals chosen for their diverse perspectives and technical expertise, in accordance with procedures approved by the National Research Council's Report Review Committee. The purpose of this independent review is to provide candid and critical comments that will assist the institution in making its published report as sound as possible and to ensure that the report meets institutional standards for objectivity, evidence, and responsiveness to the study charge. The review comments and draft manuscript remain confidential to protect the integrity of the deliberative process. We wish to thank the following individuals for their review of this report:

Emery Brown, Massachusetts General Hospital and Massachusetts Institute of Technology
Anna Gilbert, University of Michigan
Leslie Greengard, New York University
Yu-Chi Ho, Harvard University
Stephen Robinson, University of Wisconsin
Kenneth Ribet, University of California, Berkeley
Terence Tao, University of California, Los Angeles
Yannis Yortsos, University of Southern California
Bin Yu, University of California, Berkeley
Robert Zimmer, University of Chicago

Although the reviewers listed above have provided many constructive comments and suggestions, they were not asked to endorse the conclusions or recommendations nor did they see the final draft of the report before its release. The review of this report was overseen by Lawrence D. Brown of the University of Pennsylvania and C. Judson King of the University of California, Berkeley. Appointed by the National Research Council, they were responsible for making certain that an independent examination of this report was carried out in accordance with institutional procedures and that all review comments were carefully considered. Responsibility for the final content of this report rests entirely with the authoring committee and the institution.

The committee also acknowledges the valuable contribution of the following individuals, who provided input at the meetings on which this report is based or by other means:

Theodore T. Allen, Ohio State University
Yali Amit, University of Chicago
Nafees Bin Zafar, DreamWorks Animation
Emery Brown, Massachusetts General Hospital
Robert Bryant, Mathematical Sciences Research Institute
Philip Bucksbaum, Stanford University
Russel Caflisch, University of California, Los Angeles
James Carlson, Clay Mathematics Institute
William Cleveland, Purdue University
Ronald Coifman, Yale University
Peter Constantin, University of Chicago
James Crowley, Society for Industrial and Applied Mathematics
Brenda Dietrich, IBM T.J. Watson Research Center
David Donoho, Stanford University
Cynthia Dwork, Microsoft Research
Lawrence Ein, University of Illinois at Chicago
Charles Fefferman, Princeton University
Robert Fefferman, University of Chicago
John S. Gardenier, Centers for Disease Control and Prevention (ret.)
Scott Guthery, Docent Press
Alfred Hales, Institute for Defense Analyses' Center for Communications
    Research, La Jolla
Kathryn B. Hall, Hewlett Packard
James J. Higgins, Kansas State University
Shi Jin, University of Wisconsin
C. Judson King, University of California, Berkeley
William E. Kirwan, University System of Maryland
Bryna Kra, Northwestern University
Deborah Lockhart, National Science Foundation
Dana Mackenzie, mathematics writer

Wen Masters, Office of Naval Research
Donald McClure, American Mathematical Society
Jill Mesirov, Broad Institute
Diane K. Michelson, International Sematech Manufacturing Initiative
Assaf Naor, New York University
Deborah Nolan, University of California, Berkeley
Martin Nowak, Harvard University
Sastry Pantula, National Science Foundation
Colette Patt, University of California, Berkeley
Walter Polansky, Department of Energy
Adrian Raftery, University of Washington
Samuel Rankin, American Mathematical Society
Nancy Reid, University of Toronto
Fadil Santosa, University of Minnesota
Terence Sejnowski, University of California, San Diego
Harry Shum, Microsoft Corporation
James Simons, Renaissance Technologies
Douglas Simpson, University of Illinois at Urbana-Champaign
Hal Stern, University of California, Irvine
Tina Straley, Mathematical Association of America
Terence Tao, University of California, Los Angeles
Richard Taylor, Harvard University
Charles Toll, National Security Agency
Kam Tsui, University of Wisconsin
Gunther Uhlmann, University of Washington
Ron Wasserstein, American Statistical Association
S.-T. Yau, Harvard University
Bin Yu, University of California, Berkeley
Robert Zimmer, University of Chicago

Contents

Summary

OVERVIEW

The vitality of the U.S. mathematical sciences enterprise is excellent. The discipline has consistently been making major advances in research, both in fundamental theory and in high-impact applications. The discipline is displaying great unity and coherence as bridges are increasingly built between subfields of research. Historically, such bridges have served as drivers for additional accomplishments, as have the many interactions between the mathematical sciences and fields of application. Both are very promising signs. The discipline's vitality is providing clear benefits to most areas of science and engineering and to the nation.

The opening years of the twenty-first century have been remarkable ones for the mathematical sciences. The list of exciting accomplishments includes among many others surprising proofs of the long-standing Poincaré conjecture and the "fundamental lemma"; progress in quantifying the uncertainties in complex models; new methods for modeling and analyzing complex systems such as social networks and for extracting knowledge from massive amounts of data from biology, astronomy, the Internet, and elsewhere; and the development of compressed sensing. As more and more areas of science, engineering, medicine, business, and national defense rely on complex computer simulations and the analysis of expanding amounts of data, the mathematical sciences inevitably play a bigger role, because they provide the fundamental language for computational simulation and data analysis. The mathematical sciences are increasingly fundamental to the social sciences and have become integral to many emerging industries.

This major expansion in the uses of the mathematical sciences has been paralleled by a broadening in the range of mathematical science ideas and techniques being used. Much of twenty-first century science and engineering is going to be built on a mathematical science foundation, and that foundation must continue to evolve and expand.

Support for basic science is always fragile, and this may be especially true of the core mathematical sciences. In order for the whole mathematical sciences enterprise to flourish long term, the core must flourish. This requires investment by universities and by the government in the core of the subject. These investments are repaid not immediately and directly in applications but rather over the long term as the subject grows and retains its vitality. From this ever-increasing store of fundamental theoretical knowledge many innovative future applications will be drawn. To give short shrift to maintaining this store would shortchange the country.

The mathematical sciences are part of almost every aspect of everyday life. Internet search, medical imaging, computer animation, numerical weather predictions and other computer simulations, digital communications of all types, optimization in business and the military, analyses of financial risks—average citizens all benefit from the mathematical science advances that underpin these capabilities, and the list goes on and on.

> **Finding: Mathematical sciences work is becoming an increasingly integral and essential component of a growing array of areas of investigation in biology, medicine, social sciences, business, advanced design, climate, finance, advanced materials, and many more. This work involves the integration of mathematics, statistics, and computation in the broadest sense and the interplay of these areas with areas of potential application. All of these activities are crucial to economic growth, national competitiveness, and national security, and this fact should inform both the nature and scale of funding for the mathematical sciences as a whole. Education in the mathematical sciences should also reflect this new stature of the field.**

Many mathematical scientists remain unaware of the expanding role for their field, and this incognizance will limit the community's ability to produce broadly trained students and to attract more of them. A community-wide effort to rethink the mathematical sciences curriculum at universities is needed. Mechanisms to connect researchers outside the mathematical sciences with the right mathematical scientists need to be improved and more students need to be attracted to the field to meet the opportunities of the future.

> **Conclusion: The mathematical sciences have an exciting opportunity to solidify their role as a linchpin of twenty-first century research and**

technology while maintaining the strength of the core, which is a vital element of the mathematical sciences ecosystem and essential to its future. The enterprise is qualitatively different from the one that prevailed during the latter half of the twentieth century, and a different model is emerging—one of a discipline with a much broader reach and greater potential impact. The community is achieving great success within this emerging model, as recounted in this report. But the value of the mathematical sciences to the overall science and engineering enterprise and to the nation would be heightened if the number of mathematical scientists who share the following characteristics could be increased:

- They are knowledgeable across a broad range of the discipline, beyond their own area(s) of expertise;
- They communicate well with researchers in other disciplines;
- They understand the role of the mathematical sciences in the wider world of science, engineering, medicine, defense, and business; and
- They have some experience with computation.

It is by no means necessary or even desirable for all mathematical scientists to exhibit these characteristics, but the community should work toward increasing the fraction that does.

In order to move in these directions, the following will need attention:

- The culture within the mathematical sciences should evolve to encourage development of the characteristics listed in the Conclusion above.
- The education of future generations of mathematical scientists, and of all who take mathematical sciences coursework as part of their preparation for science, engineering, and teaching careers, should be reassessed in light of the emerging interplay between the mathematical sciences and many other disciplines.
- Institutions—for example, funding mechanisms and reward systems—should be adjusted to enable cross-disciplinary careers when they are appropriate.
- Expectations and reward systems in academic mathematics and statistics departments should be adjusted so as to encourage a broad view of the mathematical sciences and to reward high-quality work in any of its areas.
- Mechanisms should be created that help connect researchers outside the mathematical sciences with mathematical scientists who could be appropriate collaborators. Funding agencies and academic

departments in the mathematical sciences could play a role in lowering the barriers between researchers and brokering such connections. For academic departments, joint seminars, cross-listing of courses, cross-disciplinary postdoctoral positions, collaboration with other departments in planning courses, and courtesy appointments would be useful in moving this process forward.

- Mathematical scientists should be included more often on the panels that design and award interdisciplinary grant programs. Because so much of today's science and engineering builds on advances in the mathematical sciences, the success and even the validity of many projects depends on the early involvement of mathematical scientists.
- Funding for research in the mathematical sciences must keep pace with the opportunities.

BROADENING OF THE MATHEMATICAL SCIENCES

The mathematical sciences aim to understand the world by performing formal symbolic reasoning and computation on abstract structures. One aspect of the mathematical sciences involves unearthing and understanding deep relationships among these abstract structures. Another aspect involves capturing certain features of the world by abstract structures through the process of modeling, performing formal reasoning on the abstract structures or using them as a framework for computation, and then reconnecting back to make predictions about the world. Often, this is an iterative process. Yet another aspect is to use abstract reasoning and structures to make inferences about the world from data. This is linked to the quest to find ways to turn empirical observations into a means to classify, order, and understand reality—the basic promise of science. Through the mathematical sciences, researchers can construct a body of knowledge whose interrelations are understood and where whatever understanding one needs can be found and used. The mathematical sciences also serve as a natural conduit through which concepts, tools, and best practices can migrate from field to field.

The Committee on the Mathematical Sciences in 2025 found that the discipline is expanding and that the boundaries within the mathematical sciences are beginning to fade as ideas cross over between subfields and the discipline becomes increasingly unified. In addition, the boundaries between the mathematical sciences and other research disciplines are also eroding. Many researchers in the natural sciences, social sciences, life sciences, computer science, and engineering are at home in both their own field and the mathematical sciences. In fact, the number of such people is increasing as more and more research areas become deeply mathematical. It is easy to point to work in theoretical physics or theoretical computer science that is indistinguishable from research done by mathematicians, and similar overlap occurs with theoretical ecology, mathematical biology, bioinformatics, and an increasing number of fields.

The mathematical sciences now extend far beyond the boundaries of the institutions—academic departments, funding sources, professional societies, and principal journals—that support the heart of the field. They constitute a rich and complex ecosystem in which people who are trained in one area often make contributions in another and in which the solution to a problem in one area can emerge unexpectedly from ideas generated in another. Researchers in the mathematical sciences bring special perspectives and skills that complement those brought by mathematically sophisticated researchers with other backgrounds. And the expanding connections between the mathematical sciences and so many areas of science, engineering, medicine, and business make it ever more important to have a strong mathematical sciences community through which ideas can flow. As stated in a recent review of the mathematical sciences enterprise in the United Kingdom, "Major contributions to the health and prosperity of society arise from insights, results and algorithms created by the entire sweep of the mathematical sciences, ranging across the purest of the pure, theory inspired by applications, hands-on applications, statistics of every form, and the blend of theory and practice embodied in operational research."[1]

The committee members—like many others who have examined the mathematical sciences—believe that it is critical to consider the mathematical sciences as a unified whole. Distinctions between "core" and "applied" mathematics increasingly appear artificial; in particular, it is difficult today to find an area of mathematics that does *not* have relevance to applications. It is true that some mathematical scientists primarily prove theorems, while others primarily create and solve models, and professional reward systems need to take that into account. But any given individual might move between these modes of research, and many areas of specialization can and do include both kinds of work. The EPSRC review referenced above put this nicely:

> The contributions of the mathematical sciences community should be *considered as a whole*. Although some researchers focus some of the time on addressing real-world challenges, other researchers devise remarkable insights and results that advance and strengthen the entire discipline by pursuing self-directed adventurous research.[2]

---

[1] Engineering and Physical Sciences Research Council (EPSRC), 2010, *International Review of Mathematical Science*. EPSRC, Swindon, U.K., p. 10.

[2] Op. cit., p. 12.

Overall, the mathematical sciences share a commonality of experience and thought processes, and there is a long history of insights from one area becoming useful in another. A strong core in the mathematical sciences—consisting of basic concepts, results, and continuing exploration that can be applied in diverse ways—is essential to the overall enterprise because it serves as a common basis linking the full range of mathematical scientists.

Two major drivers of the increased reach of the mathematical sciences are the ubiquity of computational simulations—which build on concepts and tools from the mathematical sciences—and exponential increases in the amount of data available for many enterprises. The Internet, which makes these large quantities of data readily available, has magnified the impact of these drivers. Many areas of science, engineering, and industry are now concerned with building and evaluating mathematical models, exploring them computationally, and analyzing enormous amounts of observed and computed data. These activities are all inherently mathematical in nature, and there is no clear line to separate research efforts into those that are part of the mathematical sciences and those that are part of computer science or the discipline for which the modeling and analysis are performed. The health and vitality of the mathematical sciences enterprise is maximized if knowledge and people are able to flow easily throughout that large set of endeavors. The "mathematical sciences" must be defined very inclusively: The discipline encompasses a broad range of diverse activities whether or not the people carrying out the activity identify themselves specifically as mathematical scientists.

This collection of people in these interfacial areas is large. It includes statisticians who work in the geosciences, social sciences, bioinformatics, and other areas that, for historical reasons, became specialized offshoots of statistics. It includes some fraction of researchers in scientific computing and in computational science and engineering. It also includes number theorists who contribute to cryptography and real analysts and statisticians who contribute to machine learning. And it includes as well operations researchers, some computer scientists, and some physicists, chemists, ecologists, biologists, and economists who rely on sophisticated mathematical science approaches. Many of the engineers who advance mathematical models and computational simulation are also included.

Anecdotal information suggests that the number of graduate students receiving training in both mathematics and another field—from biology to engineering—has increased dramatically in recent years. If this phenomenon is as general as the committee believes it to be, it shows how mathematical sciences graduate education is contributing to science and engineering generally and also how the interest in interfaces is growing.

Recommendation: The National Science Foundation should systematically gather data on such interactions—for example, by surveying departments in the mathematical sciences for the number of enrollments in graduate courses by students from other disciplines, as well as the number of enrollments of graduate students in the mathematical sciences in courses outside the mathematical sciences. The most effective way to gather these data might be to ask the American Mathematical Society to extend its annual questionnaires to include such queries.

Program officers in the National Science Foundation's (NSF's) Division of Mathematical Sciences (DMS) and in other funding agencies are aware of the many overlaps between the mathematical sciences and other disciplines. There are many examples of flexibility in funding—mathematical scientists funded by units that primarily focus on other disciplines, and vice versa. DMS in particular works to varying degrees with other NSF units, through formal mechanisms such as shared funding programs and informal mechanisms such as program officers redirecting proposals from one division to another, divisions helping one another in identifying reviewers, and so on. For the mathematical sciences community to have a more complete understanding of its reach and to help funding agencies best target their programs, the committee recommends that a modest amount of data be collected more methodically than heretofore.

Recommendation: The National Science Foundation should assemble data about the degree to which research with a mathematical science character is supported elsewhere in that organization. (Such an analysis would be of greater value if performed at a level above DMS.) A study aimed at developing this insight with respect to statistical sciences within NSF is under way as this is written, at the request of the NSF assistant director for mathematics and physical sciences. A broader such study would help the mathematical sciences community better understand its current reach, and it could help DMS position its own portfolio to best complement other sources of support for the entire mathematical sciences enterprise. It would provide a baseline for identifying changes in that enterprise over time. Other agencies and foundations that support the mathematical sciences would benefit from a similar self-evaluation.

While the expansion of the mathematical sciences and their ever-wider reach is all to the good, the committee is concerned about the adequacy of current federal funding for the discipline in light of this expansion. The results of the two preceding Recommendations, and possibly related information when available, would make it easier to evaluate the adequacy in full detail. However, the committee does note that while the growth in federal funding for the mathematical sciences over the past decades has been strong (especially at NSF), that growth does not appear to be commensurate with the intellectual expansion found in the current study.

Conclusion: The dramatic expansion in the role of the mathematical sciences over the past 15 years has not been matched by a comparable expansion in federal funding, either in the total amount or in the diversity of sources. The discipline—especially the core areas—is still heavily dependent on the National Science Foundation.

OTHER TRENDS AFFECTING THE MATHEMATICAL SCIENCES

In addition to the growing reach of the mathematical sciences, research that is motivated by questions internal to the discipline is growing more strongly interconnected, with an increasing need for research to tap into two or more fields of the mathematical sciences. Some of the most exciting recent advances have built on fields of study—for example, probability and combinatorics—that were rarely brought together in the past. This change is nontrivial, because of the large bodies of knowledge that must be internalized by the investigator(s). Because of these interdisciplinary opportunities, education is never complete today, and in some areas older mathematicians may make more breakthroughs than in the past because so much additional knowledge is needed to work at the frontier. For these reasons, postdoctoral research training may in the future become necessary for a greater fraction of students, at least in mathematics.

Another significant change in the mathematical sciences over the past decade and more has been the establishment of additional mathematical science institutes and their greater influence on the discipline and community. These institutes now play an important role in helping mathematical scientists at various career stages learn new areas and nucleate new collaborations. Some of the institutes create linkages between the mathematical sciences and other fields, and some have important roles to play in outreach to industry and the general public. Their collective impact in changing and broadening the culture of the mathematical sciences has been enormous.

A third important trend is the rise of new modes of scholarly communication based on the Internet. While face-to-face meetings between mathematical scientists remain an essential mode of communication, it now is easy for mathematical scientists to collaborate with researchers across the world. However, new modes of collaboration and "publishing" will call for adjustments in the ways quality control is effected and professional accomplishments are measured.

The committee is concerned also about preserving the long-term accessibility of the results of mathematical research while new modes of interaction via the Internet are evolving. For example, public archives such as arXiv play a valuable role, but their long-term financial viability is far from assured and they are not used as universally as they might be. The mathematical sciences community as a whole, through its professional organizations, needs to formulate a strategy for optimizing accessibility, and NSF could take the lead in catalyzing and supporting this effort.

A final trend is the ubiquity of computing throughout science and engineering, a trend that began decades ago and escalated in the 1990s. Scientific computing has grown to be an area of study in its own right, but often it is not pursued in a unified way at academic institutions, instead existing in small clusters scattered in a variety of science and engineering departments. Mathematical sciences departments should play a role in seeing that there is a central home for computational research and education at their institutions. Beyond this, because computation is often the means by which the mathematical sciences are applied in other fields and is also the driver of many new applications of the mathematical sciences, it is important that most mathematical scientists have a basic understanding of it. Academic departments may consider seminars or other processes to make it easy for mathematical scientists to learn about the rapidly evolving frontiers of computation. Because the nature and scope of computation are continually changing, a mechanism is needed to ensure that mathematical sciences researchers have access to computing power at an appropriate scale. NSF/DMS should consider instituting programs to ensure that researchers have access to state-of-the art computing power.

PEOPLE IN THE MATHEMATICAL SCIENCES

The expansion of research opportunities in the mathematical sciences necessitates changes in the way students are prepared and a plan for how to attract more talented young people into the discipline. The demand for people with strong mathematical science skills is already growing and will probably grow even more as the range of positions that require mathematical skills expands. While these positions can often be filled by people with other postsecondary degrees, all of these individuals will need strong mathematical science skills. Because mathematical science educators have a responsibility to prepare students from many disciplines for a broad range of science, technology, engineering, and mathematics (STEM) careers, this expansion of opportunities has clear implications for the mathematical science community.

The mathematical sciences community has a critical role in educating a broad range of students. Some will exhibit a special talent in mathematics from a young age, but there are many more whose interest in the mathematical sciences arises later and perhaps through nontraditional pathways, and these latter students constitute a valuable pool of potential majors and graduate students. A third cadre consists of students from other STEM disciplines who need a strong mathematical sciences education. All three categories of students need expert guidance and mentoring from successful mathematical scientists, and their needs are not identical. The mathematical sciences must successfully attract and serve all of them.

It is critical that the mathematical sciences community actively engage with STEM discussions going on outside their own community and not be marginalized in efforts to improve STEM education, especially since the results of those efforts would greatly affect the responsibilities of mathematics and statistics faculty members. The need to create a truly compelling menu of creatively taught lower-division courses in the mathematical sciences tailored to the needs of twenty-first century students is pressing, and partnerships with mathematics-intensive disciplines in designing such courses are eminently worth pursuing. The traditional lecture-homework-exam format that often prevails in lower-division mathematics courses would benefit from a reexamination. A large and growing body of research indicates that STEM education can be substantially improved through a diversification of teaching methods. Change is unquestionably coming to lower-division undergraduate mathematics, and it is incumbent on the mathematical sciences community to ensure that it is at the center of these changes and not at the periphery.

Mathematical sciences curricula need attention. The educational offerings of typical departments in the mathematical sciences have not kept pace with the large and rapid changes in how the mathematical sciences are used in science, engineering, medicine, finance, social science, and society at large. This diversification entails a need for new courses, new majors, new programs, and new educational partnerships with those in other disciplines, both inside and outside universities. New educational pathways for training in the mathematical sciences need to be created—for students in mathematical sciences departments, for those pursuing degrees in science, medicine, engineering, business, and social science, and for those already in the workforce needing additional quantitative skills. New credentials such as professional master's degrees may be needed by those about to enter the workforce or already in it. The trend toward periodic acquisition of new job skills by those already in the workforce provides an opportunity for the mathematical sciences to serve new needs.

Most mathematics departments still tend to use calculus as the gateway to higher-level coursework, and that is not appropriate for many students. Although there is a very long history of discussion about this issue, the need for a serious reexamination is real, driven by changes in how the mathematical sciences are being used. Different pathways are needed for students who may go on to work in bioinformatics, ecology, medicine, computing, and so on. It is not enough to rearrange existing courses to create alternative curricula; a redesigned offering of courses and majors is needed. Although there are promising experiments, a community-wide effort is needed in the mathematical sciences to make its undergraduate courses more compelling to students and better aligned with the needs of user departments.

At the graduate level, many students will end up not with traditional academic jobs but with jobs where they are expected to deal with problems much less well formulated than those in the academic setting. They must bring their mathematical sciences talent and sophistication to bear on ill-posed problems so as to contribute to their solution. This suggests that graduate education in the mathematical sciences needs to be rethought in light of the changing landscape in which students may now work. At the least, mathematics and statistics departments should take steps to ensure that their graduate students have a broad and up-to-date understanding of the expansive reach of the mathematical sciences.

> **Recommendation: Mathematics and statistics departments, in concert with their university administrations, should engage in a deep rethinking of the different types of students they are attracting and wish to attract, and should identify the top priorities for educating these students. This should be done for bachelor's, master's, and Ph.D.-level curricula. In some cases, this rethinking should be carried out in consultation with faculty from other relevant disciplines.**

> **Recommendation: In order to motivate students and show the full value of the material, it is essential that educators explain to their K-12 and undergraduate students how the mathematical science topics they are teaching are used and the careers that make use of them. Modest steps in this direction could lead to greater success in attracting and retaining students in mathematical sciences courses. Graduate students should be taught about the uses of the mathematical sciences so that they can pass this information along to students when they become faculty members. Mathematical science professional societies and funding agencies should play a role in developing programs to give faculty members the tools to teach in this way.**

The community collectively does not do a good job in its interface with the general public or even with the broader scientific community, and improving this would contribute to the goal of broadening the STEM pipeline.

Recommendation: More professional mathematical scientists should become involved in explaining the nature of the mathematical sciences enterprise and its extraordinary impact on society. Academic departments should find ways to reward such work. Professional societies should expand existing efforts and work with funding entities to create an organizational structure whose goal is to publicize advances in the mathematical sciences.

The market for mathematical sciences talent is now global, and the United States is in danger of losing its global preeminence in the discipline. Other nations are aggressively recruiting U.S.-educated mathematical scientists, especially those who were born in those nations. Whereas for decades the United States has been attracting the best of the world's mathematical scientists, a reverse brain drain is now a real threat. The policy of encouraging the growth of the U.S.-born mathematical sciences talent pool should continue, but it needs to be supplemented by programs to attract and retain mathematical scientists from around the world, beginning in graduate school and continuing through an expedited visa process for those with strong credentials in the mathematical sciences who seek to establish permanent residence.

The underrepresentation of women and ethnic minorities in mathematics has been a persistent problem for the field. As white males become a smaller fraction of the population, it is even more essential that the mathematical sciences become more successful at attracting and retaining students from across the totality of the population. While there has been progress in the last 10-20 years, the fraction of women and minorities in the mathematical sciences drops with each step up the career ladder. A large number of approaches have been tried to counter this decline, and many appear to be helpful, but this problem still needs attention, and there is no quick solution.

Recommendation: Every academic department in the mathematical sciences should explicitly incorporate recruitment and retention of women and underrepresented groups into the responsibilities of the faculty members in charge of the undergraduate program, graduate program, and faculty hiring and promotion. Resources need to be provided to enable departments to adopt, monitor, and adapt successful recruiting and mentoring programs that have been pioneered at other schools and to find and correct any disincentives that may exist in the department.

While the mathematical sciences enterprise has tremendous responsibilities for educating students across the range of STEM fields, it must also, of course, replenish itself. One successful way to strengthen that part of the pipeline—of students with strong talents in the mathematical sciences per se—has been focused outreach to precollege students via mechanisms such as Math Circles.

Recommendation: The federal government should establish a national program to provide extended enrichment opportunities for students with unusual talent in the mathematical sciences. The program would fund activities to help those students develop their talents and enhance the likelihood of their pursuing careers in the mathematical sciences.

STRESSES ON THE HORIZON

Mathematical science departments, particularly those in large state universities, have a tradition of teaching service courses for nonmajors. These courses, especially the large lower-division ones, help to fund positions for mathematical scientists at all levels, but especially for junior faculty and graduate teaching assistants. But now the desire to reduce costs is pushing students to take some of their lower-division studies at state and community colleges. It is also leading university administrations to hire a second tier of adjunct instructors with greater teaching loads, reduced expectations of research productivity, and lower salaries, or to implement a series of online courses that can be taught with less ongoing faculty involvement. While these trends have been observed for a decade or more, current financial pressures may increase pressure to shift more teaching responsibilities in these ways.

The committee foresees a more difficult period for the mathematical sciences on the horizon because of this changing business model for universities. Because of their important role in teaching service courses, the mathematical sciences will be disproportionately affected by these changes. However, while there may be less demand for lower-division teaching, there may be expanded opportunities to train students from other disciplines and people already in the workforce. Mathematical scientists should work proactively—through funding agencies, university administrations, professional societies, and within their departments—to be ready for these changes.

Some educators are experimenting with lower-cost ways of providing education, such as Web-based courses that put much more burden on the students, thereby allowing individual professors to serve larger numbers of students. Some massive online open courses (MOOCs) with mathematical content have already proven to be tremendously popular, and this will only increase the interest in experimenting with this modality. While online education in the mathematical sciences is a work in progress, effective ways to deliver this material at a level of quality comparable to large university lecture classes most likely will be found. It is strongly in the interests of mathematical scientists to be involved in initiatives for online education, which will otherwise happen in a less-than-optimal way.

> Recommendation: Academic departments in mathematics and statistics should begin the process of rethinking and adapting their programs to keep pace with the evolving academic environment, and be sure they have a seat at the table as online content and other innovations in the delivery of mathematical science coursework are created. The professional societies have important roles to play in mobilizing the community in these matters, through mechanisms such as opinion articles, online discussion groups, policy monitoring, and conferences.

1

Introduction

STUDY OVERVIEW

The opening years of the twenty-first century have been remarkable ones for the mathematical sciences. Major breakthroughs have been made on fundamental research problems. The ongoing trend for the mathematical sciences to play an essential role in the physical and biological sciences, engineering, medicine, economics, finance, and social science has expanded dramatically. The mathematical sciences have become integral to many emerging industries, and the increasing technological sophistication of our armed forces has made the mathematical sciences central to national defense. A striking feature of this expansion in the uses of the mathematical sciences has been a parallel expansion in the kinds of mathematical science ideas that are being used.

There is a need to build on and solidify these gains. Too many mathematical scientists remain unaware of the expanding role for their field, and this in turn will limit the community's ability to produce broadly trained students and to attract larger numbers of students. A community-wide effort to rethink the mathematical sciences curriculum at universities is needed. Mechanisms to connect researchers outside the mathematical sciences with appropriate mathematical scientists need to be improved. The number of students now being attracted to the field is inadequate to meet the needs of the future.

A more difficult period is foreseen for the mathematical sciences because the business model for universities is entering a period of rapid change. Because of their extensive role in teaching service courses, the mathematical sciences will be disproportionately affected by these changes.

These conclusions were reached by the Committee on the Mathematical Sciences in 2025 of the National Research Council (NRC), which conducted the study that led to this report. The study was commissioned by the Division of Mathematical Sciences (DMS) of the National Science Foundation (NSF). DMS is the primary federal office that supports mathematical sciences research and the health of the mathematical sciences community. In recent years, it has provided nearly 45 percent of federal funding for mathematical sciences research and the large majority of support for research in the core areas of the discipline. Other major federal funders of the mathematical sciences include the Department of Defense, the Department of Energy, and the National Institutes of Health. Details of federal funding are given in Appendix C.

While the mathematical sciences community and its sponsors regularly hold meetings and workshops to explore emerging research areas and assess progress in more mature areas, there has been no comprehensive strategic study of the discipline since the so-called Odom study[1] in the late 1990s. During 2008, DMS Director Peter March, with encouragement from the NSF associate director for mathematical and physical sciences, Tony Chan, worked with the NRC's Board on Mathematical Sciences and Their Applications (BMSA) to define the goals of a new strategic study of the discipline.

For the study that produced this report, DMS and BMSA chose a time horizon of 2025. It was felt that a strategic assessment of the mathematical sciences needed a target date and that the date should be sufficiently far in the future to enable thinking about changes that might correspond to a generational shift. Such changes might, for example, depend on changes in graduate education that may not yet be implemented.

The specific charge for this study reads as follows:

> The study will produce a forward-looking assessment of the current state of the mathematical sciences and of emerging trends that will affect the discipline and its stakeholders as they look ahead to the quarter-century mark. Specifically, the study will assess:
>
> —The vitality of research in the mathematical sciences, looking at such aspects as the unity and coherence of research, significance of recent developments, rate of progress at the frontiers, and emerging trends;
> —The impact of research and training in the mathematical sciences on science and engineering; on industry and technology; on innovation and economic competitiveness; on national security; and other areas of national interest.

---

[1] National Science Foundation, 1998, *Report of the Senior Assessment Panel for the International Assessment of the U.S. Mathematical Sciences*. NSF, Arlington, Va.

The study will make recommendations to NSF's Division of Mathematical Sciences on how to adjust its portfolio of activities to improve the vitality and impact of the discipline.

To carry out this study, the NRC appointed a broad mix of people with expertise across the mathematical sciences, extending into related fields that rely strongly on mathematics and statistics. Biographical sketches of the committee members are included in Appendix F. As was done for two earlier NRC strategic studies chaired by former Presidential Science Advisor Edward David—*Renewing U.S. Mathematics: Critical Resource for the Future* (1984) and *Renewing U.S. Mathematics: A Plan for the 1990s* (1990)[2]—and the aforementioned one led by William Odom, a chair was sought who is not a mathematical scientist. (Dr. David was trained as an electrical engineer and General Odom was an expert on the Soviet Union.) This was done so the report would not veer into advocacy and also so it would be steered by someone with a broad view of how the mathematical sciences fit within broader academic and research endeavors. The breadth of the current committee—only half of the members sit in academic departments of mathematics or statistics—enabled the study to assess the actual and potential effects of the mathematical sciences on the broader science and engineering enterprise.

To inform its deliberations, the committee interacted with a wide range of invited speakers, as shown in Appendix B. At its first meeting, the focus was on learning about other NRC strategic studies of particular disciplines: how they were carried out and what kinds of results were produced. At that meeting, the committee also engaged in discussion with a pair of experienced university administrators, one a mathematician, to explore the changing setting for academic research and what might be on the horizon. In addition, the committee examined a large number of relevant reports and community inputs. The second meeting featured discussions with a range of individuals who employ people with mathematical skills, to explore the kinds of skills (in emerging industries, especially) that are needed, and the adequacy of the existing pipelines. Inputs from these first two meetings influenced Chapters 3 and 6 in particular.

To gather inputs from the mathematical sciences community, the committee established a Web site for input and requested comments through a mass e-mail to department heads and other community leaders, using a list maintained by DMS. It also produced announcements that were published in the May 2011 issues of the *Notices* of the American Mathematical Society (AMS) and the *AMSTAT News* of the American Statistical Associa-

---

[2] These are also known colloquially as the "David I" and "David II" reports. Both were published by the National Academy Press, Washington, D.C.

tion (ASA). Eight inputs were received through this route. The committee sent specific requests for comments to the leaders of selected committees of the AMS, the Society for Industrial and Applied Mathematics (SIAM), ASA, and the Mathematical Association of America. At its third meeting, which was held in Chicago, the committee organized a panel discussion with representatives from eight mathematical sciences departments in the vicinity of Chicago. That discussion focused on challenges and opportunities facing departments and the profession and on how to respond. Similar questions were discussed with several dozen community members at open sessions the committee held at the Joint Mathematics Meetings in New Orleans, January 2011; the International Congress of Industrial and Applied Mathematics in Vancouver, July 2011; and the Joint Statistical Meetings in Miami, August 2011. In addition, helpful discussions were held in March 2011 with the AMS Committee on Science Policy, in April 2011 with the SIAM Science Policy Committee, and in October 2011 and April 2012 with the Joint Policy Board for Mathematics.

A mechanism that proved particularly valuable was a series of 11 conference calls that members of the committee held in March-May 2011 with selected experts across the mathematical sciences. Salient observations raised by these experts (who are listed, as already mentioned, in Appendix B) are reflected in Chapters 3 and 4.

Coincidently, the current study overlapped analogous examinations in the United Kingdom and Canada. A member of the study committee chaired the U.K. assessment, and two members of the committee served on the advisory board for the Canadian assessment; in addition, the committee was briefed on the Canadian study by that study's executive director. Through these links and examination of specific materials, this study was informed by the U.K. and Canadian work.[3]

As part of this study, the committee also produced an interim product titled *Fueling Innovation and Discovery: The Mathematical Sciences in the 21st Century.* That short report highlights a dozen illustrations of research progress of recent years in a format that is accessible to the educated public and conveys the excitement of the discipline. It recounts how research in the mathematical sciences have led to Google's search algorithm, advances in medical imaging, progress in theoretical physics, technologies that contribute to national defense, methods for genomic analysis, and many other capabilities of importance to all people. But that report merely skims the surface, because the mathematical sciences nowadays touch all of us in so many ways.

---

[3] Engineering and Physical Sciences Research Council (EPSRC), 2010, *International Review of Mathematical Science.* EPSRC, Swindon, U.K.; Natural Sciences and Engineering Research Council (NSERC), 2012, *Solutions for a Complex Age: Long Range Plan for Mathematical and Statistical Sciences Research in Canada 2013–2018.* NSERC, Ottawa, Canada.

NATURE OF THE MATHEMATICAL SCIENCES

This report takes an expansive and unified view of the mathematical sciences. The mathematical sciences encompass areas often labeled as core and applied mathematics, statistics, operations research, and theoretical computer science. In the course of the study that led to this report, it became clear both that the discipline is expanding and that the boundaries within the mathematical sciences are beginning to fade as ideas cross over between subfields and the discipline becomes increasingly unified. In addition, the boundaries between the mathematical sciences and other subjects are also eroding. Many researchers in the natural sciences, social sciences, life sciences, computer science, and engineering are at home in both their own field and the mathematical sciences. In fact, the number of such people is increasing as more and more research areas become deeply mathematical. It turns out that the expansion of the mathematical sciences is a major conclusion of this report, one that is discussed in Chapter 4. The discipline has evolved considerably over the past two decades, and the mathematical sciences now extend far beyond the definitions implied by the institutions—academic departments, funding sources, professional societies, and principal journals—that support the heart of the field.

The mathematical sciences underpin a broad range of science, engineering, and technology, including the technology found in many everyday products. Many mathematical scientists are motivated by such applications, and they target their work so as to create particular mathematical and statistical understanding and capabilities. Such work—for example, the compressed sensing research highlighted in Chapter 2—usually goes far beyond routine application of an existing idea, tending to be instead very innovative and deep. A large fraction of mathematical science work is not motivated by external applications, and the reader who focused only on applications would be misled about something central to the culture of the mathematical sciences: the importance of discovery for its own sake and the quest for internal coherence, both common drivers of research. But the words (such as "beauty") that are often used to describe the motivation for such research fail to capture the power and value of the work. Whether externally or internally motivated, mathematical sciences research aims to understand deep connections and patterns. Researchers are driven to understand how the world is put together and to find its underlying order and structure. This leads to concepts with deep interconnections. When a researcher explores unanswered questions, she or he may catch glimpses of patterns, of unexpected links. The desire to understand "why" is very compelling, and this curiosity has a long history of leading to important new developments. Moreover, when a researcher succeeds in proving that those glimpses are backed up by precisely characterized connections, the way the pieces fall into place is indeed beautiful, and researchers are struck by the "rightness" or inevitability of this new insight. Synonyms for this driving concept might be "simplicity," "naturalness," "power," and "comprehensiveness," and mathematical scientists of all stripes put a premium on results with intellectual depth, generality, and an ability to explain many things at once and to expose previously hidden interconnections (integrating ideas from disparate areas).

Even when research is internally motivated, it is strikingly common to find instances in which applications arise in a different discipline and the necessary mathematics is already available, having been generated by mathematical scientists for unrelated reasons. As one example, the committee cites an interchange in the early 1970s between the mathematician Jim Simons and the theoretical physicist Frank Yang, when Yang was explaining a theory he was trying to develop to help him understand elementary particles in physics. Simons—whose background in mathematics later gave him the foundation for a very successful career shift into finance—said to Yang, "Stop, don't do that." Yang, taken aback, asked, "Why not?" Simons said, "Because mathematicians already did it more than 30 years ago." Yang then asked, "For what reason? Why would they ever do that?" The answer is of course that they were motivated by the internal, esthetic considerations of their, at the time, completely theoretic investigations. This is not an isolated incident, but rather an example of what happens repeatedly in the mathematical sciences. The prime numbers and their factorization, initially studied for esthetic reasons, now provide the underpinnings of Internet commerce. Riemann's notion of geometry and curvature later became the basis of Einstein's general relativity. Quaternions, whose multiplication table was triumphantly carved into a Dublin bridge by William Hamilton in 1843, are now used in video games and in tracking satellites. Operators on Hilbert space provided the natural framework for quantum mechanics. Eigenvectors are the basis for Google's famous Page Rank algorithm and for software that recommends other products to users of services such as Netflix. Integral geometry makes possible MRI and PET scans. The list of such examples is limited only by the space to tell about them. Some additional examples are given in Chapter 2, where the interplay between theoretical physics and geometry is described.

A strong core in the mathematical sciences—consisting of basic concepts, results, and continuing exploration that can be applied in diverse ways—is essential to the overall enterprise because it serves as a common basis linking the full range of mathematical scientists. Researchers in far-flung specialties can find common language and link their work back to common principles. Because of this, there is a coherence and interdependence across the entire mathematical sciences enterprise, stretching from the most theoretical to the most applied.

Robert Zimmer, a mathematical scientist who is president of the University of Chicago, speaks of the mathematical sciences as a fabric: If it is healthy—strong and connected throughout the whole—then it can be tailored and woven in many ways; if it has disconnects, then its usefulness has limitations. He also argues that, because of this interconnectivity, there is a degree of inevitability to the ultimate usefulness of mathematical sciences research. That is, important applications are the rule rather than the exception.[4] Over and over, research that was internally motivated has become the foundation for applied work and underlies new technologies and start-ups. And often questions that arise because of our inability to mathematically represent important phenomena from applications prompt mathematical scientists to delve back into fundamental questions and create additional scaffolding of value both to the core and to future applications.

The fabric metaphor accurately captures the interconnectivity of the various strands of the mathematical sciences; all of the strands are woven together, each supporting the others, and collectively forming an integrated whole that is much stronger than the parts separately. The mathematical sciences function as a complex ecosystem. Ideas and techniques move back and forth—innovations at the core radiate out into applied areas; flowing back, new mathematical problems and concepts are drawn forth from problems arising in applications. The same is true of people—those who choose to make their careers in applied areas frequently got a significant part of their training from core mathematical scientists; seeing the uses and power of mathematics draws some people in to study the core. One never knows from which part of the mathematical sciences the next applications will come, and one never knows whether what is needed for a possible application is existing knowledge, a variation on what already exists, or something completely new. To maintain U.S. leadership in the mathematical sciences, the entire ecosystem must remain healthy.

EVERYONE SHOULD CARE ABOUT THE
MATHEMATICAL SCIENCES

In everyday life, terms that sound mathematical increasingly appear in a variety of contexts. "Doing the math" is used by politicians to mean analyzing the gains or losses of doing something, and language such as "exponential," "algorithm," and "in the equation" frequently appears in business and finance. A positive interpretation of this phenomenon is that more and more people appreciate the mathematical sciences, but a not-so-

---

[4] For this reason, this report tends to avoid the terms "core mathematics" and "applied mathematics." As can be seen in many places in the report, nearly all areas of the mathematical sciences can have applications.

fortunate consequence is that the average person may not appreciate the richness of the mathematical sciences.

The mathematical sciences include far more than numbers—they deal with geometrical figures, logical patterns, networks, randomness, and predictions from incomplete data, to name only a few topics. And the mathematical sciences are part of almost every aspect of everyday life.

Consider a typical man (Bob) and a typical woman (Alice) in a developed society such as the United States. Whether they know it or not, their lives depend intimately and deeply on the mathematical sciences; they are wrapped in an intricate and elegant net woven with strands from the mathematical sciences. Here are some examples. A remarkable fact is that these extremely varied applications depend crucially on the body of mathematical theory that has been developed over hundreds of years—on ingenious new uses of theoretical developments from long ago, but also on some very recent breakthroughs. Some of the pioneers of this body of theory were motivated by these applications; some by other applications that would seem completely unconnected with these; and in many cases by the pure desire to explore the fundamental structures of science and thought.

- Bob is awakened by a radio clock and usually listens to the news. But he is unlikely to think twice, if at all, about how the radio can receive signals, remove noises, and produce pleasant sound, yet all of these tasks involve the mathematical and statistical methods of signal processing.

- Alice may begin her day by watching the news in a recently purchased high-definition LCD television. To achieve the high-quality image that Alice takes for granted, many sophisticated steps are required that depend on the mathematical sciences: compression of digital signals, conversion from digital to analog and analog to digital, image analysis and enhancement, and LCD performance optimization.

- Bob and Alice love movies like *Toy Story, Avatar,* and *Terminator 3.* A growing number of films feature characters and action scenes that are the result of calculations performed by computers on mathematical models of movements, expressions, and actions based on mathematical models. Obtaining a realistic impression of, say, the collapse of downtown Los Angeles, requires intricate mathematical characterizations of explosions and their aftermath, displayed through the application of high-end computational power to sophisticated mathematical insights about the fundamental equations governing fluids, solids, and heat.

- If Bob's plans for his day (or the next few days) take into account weather predictions, he is relying on the numerical solution of highly nonlinear, high-dimensional (meaning tens of millions of unknowns) equations and on statistical analysis of past observations integrated with freshly collected information about atmosphere and ocean conditions.

- To surf the Internet, Bob turns to a search engine, which performs rapid searches using a sophisticated mathematical algorithm. The earliest Web search techniques treated the interconnections of the Web as a matrix (a two-dimensional data array), but modern search methods have become much more sophisticated, incorporating protection from hacking and manipulation by outsiders. Effective Web search relies more than ever on sophisticated strategies derived from the mathematical sciences.

- Alice is plagued by unwanted e-mails from irrelevant people who want to cheat her or sell her items she does not want. A common solution to this problem is a spam filter, which tries to detect unwanted or fraudulent e-mail using information and probability theory. A major underlying tool is machine learning, in which features of "legitimate" e-mails (as assessed by humans) are used to train an algorithm that classifies incoming e-mails as legitimate or as spam.

- When Alice needs to attend a meeting next month in Shanghai, China, both the schedule of available flights and the price she will pay for her ticket are almost certain to be determined using optimization (by the airlines).

- Bob uses his cell phone almost constantly—a feature of modern life enabled, for better or worse, by new developments in mathematical and statistical information theory that involve wireless signal encoding, transmission, and processing, and by some highly ingenious algorithms that route the calls.

- Alice's office building consumes energy to run electric lights, telephone landlines, a local computer network, running water, heating, and cooling. Mathematical optimization and statistical techniques are used to plan for efficient energy delivery based on information about expected energy consumption and estimated safety factors to protect against unusual events such as power outages. Because of concerns about excessive energy usage, Alice's utility companies are investing in new mathematical and statistical techniques for planning, monitoring, and controlling future energy systems.

- When Bob and Alice need medical or dental attention, they explicitly benefit from sophisticated applications of the mathematical sciences. Everyone has heard of X-rays, CT scans, and MRIs, but few people realize that modern medical and dental image analysis and interpretation depend on complicated mathematical concepts, such as the Radon and Fourier transforms, whose theory was initially developed during the nineteenth century. This example illustrates the crucial observation that research in the mathematical sciences has a very long shelf life in the sense that, because of their abstract nature, discoveries in the mathematical sciences do not become obsolete. Hence, a fresh insight about their application may arise several decades (or more) after their publication.

- When Alice's doctor prescribes a new medication, she depends on decisions by pharmaceutical companies and the government about the effectiveness and safety of new drugs and chemical treatments— and those decisions in turn depend on ever-improving statistical and mathematical methods. Companies use mathematical models to predict how possible new drug molecules are expected to interact with the body or its invaders and combinatorial and statistical methods to explore the range of promising permutations.

- When Alice and Bob order products online, the processes used for inventory management and control, delivery scheduling, and pricing involve ingredients from the mathematical sciences such as random matrices, scheduling and optimization algorithms, decision theory, statistical regression, and machine learning.

- If Alice or Bob borrows money for a house, car, education, or to pay off a credit card or invest savings in stocks, bonds, real estate, mutual funds, or their 401(k)s, the mathematical sciences are hard at work in the financial markets and the related micro- and macroeconomics. Mathematical methods, statistical projections, and computer modeling based on data are all necessary to function, prosper, and plan for daily life and retirement in today's array of global markets. Today, many tools are brought directly to individuals in customized applications for virtually every type of personal computer and personal communication device.

- When Alice enters an airport or bus station, surveillance cameras are likely to record her movements as well as those of everybody in the area. The enormous task of processing and assessing the images from multiple closed-circuit recordings is sometimes done by mathematical tools that automatically analyze movement patterns to determine which people are likely to be carrying hidden weapons or explosives. Similar techniques are being applied in stores and shopping centers to assess which people are likely to be shoplifters or thieves.

- Even when Bob stops at the supermarket on the way home, he cannot escape the mathematical sciences, which are used by retailers to place products in the most appealing locations, to give him a set of discount coupons chosen based on his past shopping history, and to price items so that total sales revenue will be maximized.

BOX 1-1
Four Facts Most People Don't Know
About the Mathematical Sciences

Mathematical scientists have varied careers and styles of work. They do not spend all their time calculating—though some do quite a bit—nor do most of them toil in isolation on abstract theories. Most engage in collaborations of some sort. While the majority are professors, there are also many mathematical scientists in pharmaceutical and manufacturing industries, in government and national defense laboratories, in computing and Internet-based businesses, and on Wall Street. Some mathematicians prove theorems, but many others engage in other aspects of quantitative modeling and problem solving. Mathematical scientists contribute to every field of science, engineering, and medicine.

The mathematical sciences are always innovating. They do not consist of a fixed collection of facts that are learned once and thereafter simply applied. While theorems, once proved, may continue to be useful on a time-scale of centuries, new theorems are constantly being discovered, and adapting existing knowledge to new contexts is a never-ending process.

The United States is very good at the mathematical sciences. In spite of concerns about the average skill of precollege students, the United States has an admirable record of attracting the best mathematical and statistical talent to its universities, and many of those people make their homes here after graduation. Assessments of capabilities in mathematical sciences research find the United States to be at or near the top in all areas of the discipline.

Mathematical scientists can change course during their careers. Because the mathematical sciences deal with methods and general principles, researchers need not maintain the same focus for their entire careers. For example, a statistician might work on medical topics, climate models, and financial engineering in the course of one career. A mathematician might find that insights from research in geometry are also helpful in a materials science problem, or in a research challenge from brain imaging. And new types of mathematical science jobs are constantly being created.

This chapter concludes with Box 1-1, "Four Facts Most People Don't Know About the Mathematical Sciences," which illustrates some attributes of today's mathematical sciences.

STRUCTURE OF THE REPORT

Chapter 2 discusses recent accomplishments of the mathematical sciences and the general health of the discipline. While the situation is very good at present, stresses and challenges are on the horizon. Chapter 3 summarizes the current state of the mathematical sciences. Chapter 4 draws from the inputs to the study and from committee members' own experiences to identify trends that are affecting the mathematical sciences. It also identifies emerging stresses and challenges. Chapter 5 discusses the pipeline that prepares people for mathematical science careers, while Chapter 6 discusses the ramifications of emerging changes in the academic environment.

2

Vitality of the Mathematical Sciences

The vitality of the U.S. mathematical sciences enterprise is excellent. The mathematical sciences have consistently been making major advances in both fundamental theory and high-impact applications, and recent decades have seen tremendous innovation and productivity. The discipline is displaying great unity and coherence as more and more bridges are built between subfields of research. Historically, such bridges serve as drivers for additional accomplishments, as do the many interactions between the mathematical sciences and fields of application, and so the existence of several striking examples of bridge-building in this chapter is a very promising sign for the future. The programs of the National Science Foundation's (NSF) mathematical science institutes offer good evidence of this bridge-building, and the large-scale involvement of graduate students and postdoctoral researchers at those institutes suggests that the trend will continue. New tools for scholarly communications, such as blogs and open-access repositories of research, contribute to the excellent rate of progress at the research frontiers. As shown in this chapter and in the separate report *Fueling Innovation and Discovery: The Mathematical Sciences in the 21st Century*, the discipline's vitality is providing clear benefits for diverse areas of science and engineering, for industry and technology, for innovation and economic competitiveness, and for national security.

Further down the road, toward 2025, stresses are likely, and these will be discussed in Chapters 4 through 6. The focus of this chapter is to document some recent advances as illustrations of the health and vitality of the mathematical sciences. Indeed, the growth of new ideas and applications in the mathematical sciences is so robust that inevitably it far exceeds the spectrum of expertise of a modest-sized committee. What appears below is just a sampling of advances, to give a flavor of what is going on, and it is not meant to be either comprehensive or proportionally representative. This chapter is aimed primarily at the mathematical sciences community, and so the examples here presume a base of mathematical or statistical knowledge. The topics covered range from the solution of a century-old problem by using techniques from one field of mathematics to solve a major problem in another, to the creation of what are essentially entirely new fields of study. The topics are, in order:

- The Topology of Three-Dimensional Spaces
- Uncertainty Quantification
- The Mathematical Sciences and Social Networks
- The Protein-Folding Problem and Computational Biology
- The Fundamental Lemma
- Primes in Arithmetic Progression
- Hierarchical Modeling
- Algorithms and Complexity
- Inverse Problems: Visibility and Invisibility
- The Interplay of Geometry and Theoretical Physics
- New Frontiers in Statistical Inference
- Economics and Business: Mechanism Design
- Mathematical Sciences and Medicine
- Compressed Sensing

THE TOPOLOGY OF THREE-DIMENSIONAL SPACES

The modest title of this section hides a tremendous accomplishment. The notion of space is central to the mathematical sciences, to the physical sciences, and to engineering. There are entire branches of theoretical mathematics devoted to studying spaces, with different branches focusing on different aspects of spaces or on spaces endowed with different characteristics or structures.[1] For example, in topology one studies spaces without assuming any structure beyond the notion of coherence or continuity. By contrast, in geometry one studies spaces in which, first of all, one can differentiate, leading to notions such as tangent vectors, and, second, for which the notion of lengths and angles of tangent vectors are defined. These concepts were first introduced by Riemann in the 1860s in his thesis "The hypotheses that underlie geometry," and the resulting structure is called a Riemannian metric. Intuitively, one can imagine that to a topologist spaces are made out of rubber or a substance like taffy, while

---

[1] Box 2-1 discusses the concept of mathematical structures.

BOX 2-1
Mathematical Structures

At various points, this chapter refers to "mathematical structures." A mathematical structure is a mental construct that satisfies a collection of explicit formal rules on which mathematical reasoning can be carried out. An example is a "group," which consists of a set and a procedure by which any two elements of the set can be combined ("multiplied") to give another element of the set, the product of the two elements. The rules a group must satisfy are few: the existence of an identity element, of inverses for each element of the set, and the associative property for the combining action. Basic arithmetic conforms with this definition: For example, addition of integers can be described in these terms. But the concept of groups is also a fundamental tool for characterizing symmetries, such as in crystallography and theoretical physics. This level of abstraction is helpful in two important ways: (1) it enables precise examinations of mathematical sets and operations by stripping away unessential details and (2) it opens the door to logical extensions from the familiar. As an example of the latter benefit, note that the definition of a group allows the combination of two elements to depend on the order in which they are "multiplied," which is contrary to the rules of arithmetic. With the explicit recognition that that property is an assumption, not a necessary consequence, mathematicians were able to define and explore "noncommutative groups" for which the order of "multiplication" *is* significant. It turns out that there are many natural situations that are naturally represented by a noncommutative group.

It is possible, of course, to define a mathematical structure that is uninteresting and that has no relevance to the real world. What is remarkable is how many interesting mathematical structures there are, how diverse are their characteristics, and how many of them turn out to be important in understanding the real world, often in unanticipated ways. Indeed, one of the reasons for the limitless possibilities of the mathematical sciences is the vast realm of possibilities for mathematical structures. Complex numbers, a mathematical structure build around the square root of -1, turn out to be rooted in the real world as part of the essential equations describing electromagnetism and quantum theory. Riemannian metrics, the mathematical structure developed to describe objects whose geometry varies from point to point, turns out to be the basis for Einstein's description of gravitation. "Graphs" (these are not the graphs used to plot functions in high school) consisting of "nodes" joined by "edges," turn out to be a fundamental tool used by social scientists to understand social networks.

A striking feature of mathematical structures is their hierarchical nature—it is possible to use existing mathematical structures as a foundation on which to build new mathematical structures. For example, although the meaning of "probability" has long vexed philosophers, it has been possible to create a mathematical structure called a "probability space" that provides a foundation on which realistic structures can be built. On top of the structure of a probability space, mathematical scientists have built the concept of a random variable, which encapsulates in a rigorous way the notion of a quantity that takes its values according to a

continued

BOX 2-1 Continued

certain set of probabilities, such as the roll of a pair of dice—one gets a different random variable depending on whether the dice are honest or loaded. There then are certain broad classes of random variables, of which the most famous are Gaussian random variables, the source of the well-known bell curve and which provide the rigorous basis of many of the fundamental tools of statistics. These different classes of random variables can be put together into structures known as probabilistic models, which are an incredibly flexible class of mathematical structures used to understand phenomena as diverse as what goes on inside a cell, financial markets, or the physics of superconductors.

Mathematical structures provide a unifying thread weaving through and uniting the mathematical sciences. For example, algorithms represent a class of mathematical structure; algorithms are often based on other mathematical structures, such as the graphs mentioned earlier, and their effectiveness might be measured using probabilistic models. Partial differential equations (PDEs) are a class of mathematical structure built on the most basic of mathematical structures— functions. Most of the fundamental equations of physics are described by PDEs, but the mathematical structure of PDEs can be studied rigorously independent of knowing, say, what the ultimate structure of space looks like at subnuclear scales. Such study can, for example, provide insight into the structures of the potential solutions of a certain PDE, explaining which phenomenology can be captured by that PDE and which cannot. Computations involving PDEs involve yet another type of mathematical structure, a "discretization scheme," which transforms what is a fundamentally continuous problem into one involving just a very large but finite set of values. The catch is that finding the right discretization scheme is highly subtle, and discretization is one example of the mathematical depth of the computational wing of the mathematical sciences.

---

to a geometer they are made out of steel. Although we have no direct visual representation of spaces of higher dimension, they exist as mathematical objects on the same footing as lower-dimensional spaces that we can directly see, and this extension from physical space has proved very useful. Topological and geometric spaces are central objects in the mathematical sciences. They are also ubiquitous in the physical sciences, in computing, and in engineering, where they are the context in which problems are precisely formulated and results are expressed.

Over 100 years ago Poincaré initiated the abstract and theoretical study of these higher-dimensional spaces and posed a problem about the three-dimensional sphere (a unit sphere in a four-coordinate space whose surface has three dimensions) that motivated topological investigations into three-dimensional spaces over the century that followed.[2] The three-dimensional sphere sits in ordinary four-dimensional coordinate space as the collection of vectors of unit length: that is, the space

$$\{(x,y,z,w) \mid x^2 + y^2 + z^2 + w^2 = 1\}$$

It has the property that any closed path (one that starts and ends at the same point) on the sphere can be continuously deformed through closed paths, always staying on the sphere, so as to shrink it down to a point path—that is, a path in which no motion occurs. Poincaré asked if this was the only three-dimensional space of finite extent, up to topological equivalence, with this property. For the next 100 years this problem and its generalizations led to enormous theoretical advances in the understanding of three-dimensional spaces and of higher-dimensional spaces, but Poincaré's original problem remained unsolved. The problem proved so difficult, and the work it stimulated so important, that in 2000 the Clay Mathematics Institute listed it as one of its seven Millennium Prize problems in mathematics, problems felt to be among the hardest and most important in theoretical mathematics.

This is purely a problem in topology. The condition about paths is on the face of it a topological condition (since to make sense of it one only needs the notion of continuity), and the conclusion is explicitly topological. For nearly 100 years it was attacked in purely topological terms. Then in 2002, Grigory Perelman succeeded in answering this question in the affirmative. Resolving a central question that has been the focus of attention for many decades is an exciting event. Here the excitement was accentuated by the fact that the solution required powerful results from other parts of theoretical mathematics, suggesting connections between them that had not been suspected. Perelman invoked deep ideas from analysis and parabolic evolution equations developed by Richard Hamilton. Briefly, Hamilton had introduced and studied an evolution equation for Riemannian metrics that is an analogue of the heat equation. Perelman was able to show that under the topological hypotheses of Poincaré's conjecture, Hamilton's flow converged to the usual metric on the 3-sphere, so that the underlying topological space is indeed topologically equivalent to the 3-sphere. But Perelman's results applied to Riemannian metrics on every three-dimensional space and gave a description of any such space in terms of simple geometric pieces, which is exactly what William Thurston had conjectured some 20 years earlier. Ironically, the ideas that led Thurston to his conjectured description

---

[2] The convention in mathematics is that the surface of a basketball is a 2-sphere (because it has two dimensions) even though it sits in three-dimensional space. The entire basketball including the air inside is called a "ball" or a "3-ball" rather than a sphere.

have their source in other, more geometric, parts of Poincaré's work, his work on Fuchsian and Kleinian groups.

Perelman's solution brings to a close a chapter of theoretical mathematics that took approximately 100 years to write, but at the same time it has opened up new avenues for the theoretical study of space. This mathematical breakthrough is fairly recent and it is too early to accurately assess its full impact inside both mathematics and the physical sciences and engineering. Nevertheless, we can make some educated guesses. Even though this is the most abstract and theoretical type of mathematics, it may well have practical import because spaces are so prevalent in science and engineering. What Perelman achieved for the particular evolution equation that he was studying was to understand how singularities develop as time passes. That particular equation is part of a general class of equations, with the heat equation being the simplest. In mathematics, various geometric problems belong to this class, as do equations for the evolution of many different types of systems in science and engineering. Understanding singularity development for these equations would have a huge impact on mathematics, science, and engineering, because the behavior of solutions near singularities can be so important. Already, we are seeing the use of the techniques Perelman introduced to increase the understanding in other geometric contexts, for example in complex geometry. Time will tell how far these ideas can be pushed, but if they extend to equations in other more applied contexts, then the payoff in practical terms could well be enormous.

UNCERTAINTY QUANTIFICATION

Central to much of science, engineering, and society today is the building of mathematical models to represent complex processes. For example, aircraft and automotive manufacturers routinely use mathematical representations of their vehicles (or vehicle components) as surrogates for building physical prototypes during vehicle design, relying instead on computer simulations that are based on those mathematical models. The economic benefit is clear. A prototype automobile that is destroyed in a crash test, for example, can cost $300,000, and many such prototypes are typically needed in a testing program, whereas a computer model of the automobile that can be virtually crashed under many varying conditions can often be developed at a fraction of the cost.

The mathematical modeling and computational science that underlies the development of such simulators of processes has seen amazing advances over the last two decades, and improvements continue to be made. Yet the usefulness of such simulators is limited unless it can be shown that they are accurate in predicting the real process they are simulating.

A host of issues arise in the process of ensuring that the simulators are accurate representations of the real process. The first is that there are typically many elements of the mathematical model (such as rate coefficients) that are unknown. Also, inputs to the simulators are often themselves imperfect—for example, weather and climate predictions must be initiated with data about the current state, which is not known completely. Furthermore, the modeling is often done with incomplete scientific knowledge, such as incomplete representation of all the relevant physics or biology, and approximations are made because of computational needs—for example, weather forecasting models being run over grids with 100-kilometer edges, which cannot directly represent fine-scale behaviors, and continuous equations being approximated by discrete analogues.

The developing field for dealing with these issues is called Uncertainty Quantification (UQ), and it is crucial to bringing to fruition the dream of being able to accurately model and predict real complex processes through computational simulators. Addressing this problem requires a variety of mathematical and statistical research, drawing on probability, measure theory, functional analysis, differential equations, graph and network theory, approximation theory, ergodic theory, stochastic processes, time series, classical inference, Bayesian analysis, importance sampling, nonparametric techniques, rare and extreme event analysis, multivariate techniques, and so on.

UQ research is inherently interdisciplinary, in that disciplinary knowledge of the process being modeled is essential in building the simulator, and disciplinary knowledge of the nature of the relevant data is essential in implementing UQ. Thus, effective UQ typically requires interdisciplinary teams, consisting of disciplinary scientists plus mathematicians and statisticians.

This appreciation for the need for UQ has grown into a clamor; calls for investment in UQ research can be found in any number of science and engineering advisory reports to funding agencies. Good progress has been made, and building the necessary capabilities for UQ is essential to the reliable use of computational simulation. The clamor has also been noticed by the mathematical and statistical societies. Interest groups in UQ have been started by the Society for Industrial and Applied Mathematics and the American Statistical Association. The two societies have also started a joint journal, *Uncertainty Quantification*.

THE MATHEMATICAL SCIENCES AND SOCIAL NETWORKS

The emergence of online social networks is changing behavior in many contexts, allowing decentralized interaction among larger groups and fewer geographic constraints. The structure and complexity of these networks have grown rapidly in recent years. At the same time, online networks are collecting social data at unprecedented scale and resolution, making social networks visible that in the past could only be explored via in-depth surveys. Today, millions of people leave digital traces of their personal social networks in the form of text messages on mobile phones, updates on Facebook or Twitter, and so on.

The mathematical analysis of networks is one of the great success stories of applying the mathematical sciences to an engineered system, going back to the days when AT&T networks were designed and operated based on graph theory, probability, statistics, discrete mathematics, and optimization. However, since the rise of the Internet and social networks, the underlying assumptions in the analysis of networks have changed dramatically. The abundance of such social network data, and the increasing complexity of social networks, is changing the face of research on social networks. These changes present both opportunities and challenges for mathematical and statistical modeling.

One example of how the mathematical sciences have contributed to the new opportunity is the significant amount of recent work focused on the development of random-graph models that capture some of the qualitative properties observed in large-scale network data. The mathematical models developed help us understand many attributes of social networks. One such attribute is the degree of connectivity of a network, which in some cases reveals the smallness of world, in which very distant parts of a population are connected via surprisingly short paths.[3] These short paths are surprisingly easy to find, which has led to the success of decentralized search algorithms.

Another important direction is the development of mathematical models of contagion and network processes. Social networks play a fundamental role in spreading information, ideas, and influence. Such contagion of behavior can be beneficial when a positive behavior change spreads from person to person, but it can also produce negative outcomes, as in cascading failures in financial markets. Such concepts open the way to epidemiological models that are more realistic than the "bin" models that do not take the structure of interpersonal contacts into account. The level of complexity in influencing and understanding such contagion phenomena rises with the size and complexity of the social network. Mathematical models have great potential to improve our understanding of these phenomena and to inform policy discussions aimed at enhancing system performance.

---

[3]See http://www.nytimes.com/2011/11/22/technology/between-you-and-me-4-74-degrees.html?_r=2&hp.

THE PROTEIN-FOLDING PROBLEM

Knowing the shape of a protein is an essential step in understanding its biological function. In Nobel-prize winning work, biologist Christian Anfinsen showed that an unfolded protein could refold spontaneously to its original biologically active conformation. This observation led to the famous conjecture that the one-dimensional sequence of amino acids of a protein uniquely determines the protein's three-dimensional structure. That in turn led to the almost 40-year effort of quantitative scientists in searching for computational strategies and algorithms for solving the so-called "protein-folding problem," which is to predict a protein's three-dimensional structure from its primary sequence information. Some subproblems include how a native structure results from the interatomic forces of the sequence of amino acids and how a protein can fold so fast.[4] Although the protein-folding conjecture has been shown to be incorrect for a certain class of proteins—for example, sometimes enzymes called "chaperones" are needed to assist in the folding of a protein—scientists have observed that more than 70 percent of the proteins in nature still fold spontaneously, each into its unique three-dimensional shape.

In 2005, the protein-folding problem was listed by *Science* magazine as one of the 125 grand unsolved scientific challenges. The impact of solving the protein-folding problem is enormous. It will have a direct and profound impact on our understanding of life: how these basic units, which carry out almost every function in living cells, play their roles at the fundamental physics level. Everything about the molecular mechanism of protein motions—folding, conformational changes, and evolution—could be revealed, and the whole cell could be realistically modeled. Furthermore, such an advance would have a great influence on novel protein design and rational drug design, which may revolutionize the pharmaceutical industry. For example, drugs might be rather accurately designed on a computer without much experimentation. Genetic engineering for improving the function of particular proteins (and thus certain phenotypes of an individual) could become realistic.

Conceptually, the protein-folding problem is straightforward: Given the positions of all atoms in a protein (typically tens of thousands of atoms), one would calculate the potential energy of the structure and then find a configuration that minimizes that energy. However, such a goal is technically difficult to achieve owing to the extreme complexity of the means by which the energy depends on the structure. A more attractive strategy, termed "molecular dynamics," has a clear physical basis: One uses

---

[4] Dill et al., 2007, The protein folding problem: When will it be solved? *Current Opinion in Structural Biology* 17: 342-346.

Newton's law of motion to write down a set of differential equations, called Hamiltonian equations, to describe the locations and speeds of all the atoms involved in the protein structure at any instant. Then, one can then numerically solve the protein structural motion equations, which not only predict low-energy structures of a protein but also provide information on protein movements and dynamics. To achieve the goal, one typically discretizes the time and approximates the differential equations by difference equations. Then, a molecular dynamic algorithm such as the leapfrog algorithm is used to integrate over the equations of motion. However, because the system is large and complex, the time step in the discretization must be sufficiently small to avoid disastrous conflicts, which implies that the computation cost is extremely high for simulating even the tiny fraction of a second that must be simulated.

Another strategy is based on the fundamental principle of statistical mechanics, which states that the probability of observing a particular structural state is proportional to its Boltzmann distribution, which is of the form $P(S) \propto e^{-E(S)/kT}$, where $E(S)$ is the potential energy of structural state S. Thus, one can use Monte Carlo methods to simulate the structural state S from this distribution. However, because of the extremely high dimensionality and complexity of the configuration space and also because of the complex energy landscape, simulating well from the Boltzmann distribution is still very challenging. New Monte Carlo techniques are needed to make the simulation more efficient, and these new methods could have a broader impact on other areas of computation as well.

Both molecular dynamics and Monte Carlo methods rely on a good energy function $E(S)$. Although much effort has been made and many insights have been gained, it is still an important challenge to accurately model interatomic interactions, especially in realistic environments—for example, immersed in water or attached to a membrane—in a practical way. The overall approach is still not as precise as would be desired. Given the availability of a large number of solved protein structures, there is still room for applying certain statistical learning strategies to combine information from empirical data and physics principles to further improve the energy function.

In recent years, great successes have been made in computational protein structure prediction by taking advantage of the fast-growing protein structure databases. A well-known successful strategy is called "homology modeling," or template-based modeling, which can provide a good approximate fold for a protein that has a homologous "relative"—that is, one with sequence similarity >30 percent—whose structure has been solved. Another attractive strategy successfully combines empirical knowledge of protein structures with Monte Carlo strategies for simulating protein folds. The idea is to implement only those structural modifications that are supported by observed structural folds in the database. So far, these bioinformatics-based learning methods are able to predict well the folding of small globular proteins and those proteins homologous to proteins of known structure.

Future challenges include the prediction of structures of medium to large proteins, multidomain proteins, and transmembrane proteins. For multidomain proteins, many statistical learning strategies have been developed to predict which domains tend to interact with one another based on massive amounts of genomic and proteomic data. More developments in conjunction with structure modeling will be needed. There are other more technical, but also very important, challenges, such as how to rigorously evaluate a new energy function or a sampling method; how to more objectively represent a protein's structure as an ensemble of structures instead of a single representative; how to evaluate the role of entropies; and how to further push the frontier of protein design. The potential impact of solving the protein-folding problem in its current form is muted by a grander, much more challenging issue. That is, how to come up with a description of protein folding and structure prediction, as well as of function and mechanism, that accords with a quantum mechanical view of reality. A protein's dynamic properties, given that they presumably conform to the laws of quantum electrodynamics, may exhibit unexpected, counterintuitive behavior unlike anything that we have ever seen or that can be predicted based on classical physics, which is the prevailing viewpoint now owing to computational limitations. For example, when a protein folds, does it utilize "quantum tunneling" to get around (imaginary) classical energy barriers that would cause problems for a molecular dynamics program? This is a mathematical and algorithmic problem that has been largely left to one side by biologists because it appears to be intractable. Some highly novel, outside-the-box mathematical concepts will likely be required if one is to overcome these limitations. An important challenge for mathematical biologists is to help discover additional protein properties, beyond the handful identified thus far, that are nonclassical and thus counterintuitive. Currently, statistical approaches have provided an indirect way of attacking such problems and a way that is analogous to the statistical approaches used by classical geneticists to work around their lack of molecular biological and cytological methods. In a similar way, statistical approaches, applied in an evolutionary context, may augment our current arsenal of experimental and theoretical methods for understanding protein folding and predicting protein structure, function, and mechanisms.

THE FUNDAMENTAL LEMMA

The fundamental lemma is a seemingly obscure combinatorial identity introduced by Robert Langlands in 1979, as a component of what is now universally called the Langlands program. The program as a whole consists of a set of conjectures that, addressed methodically, could resolve the most fundamental questions of number theory, phrased in the language of automorphic representations of the absolute Galois group.

The fundamental lemma is a statement about symmetries defined by objects called algebraic groups. One of several families of such groups is the group of linear transformations of an n-dimensional vector space. Some idea of the sweeping scope of the lemma and the Langlands program that rests on it can be gleaned from the fact that the proof of Fermat's Last Theorem, a problem that lay open for over 300 years, rested on just the case $n = 2$.[5] Langlands initially believed that the fundamental lemma would be an easy first step—although even its statement requires too much specialized knowledge and notation to accomplish here—and he assigned it as a thesis problem for a student, Robert Kottwitz. Through the work of Kottwitz and many others, some special cases were proven, but the general case remained unproven year after year. The lack of a proof came to be seen as the great obstacle to carrying out the program.

The fundamental lemma stimulated 30 years of valuable research in representation theory, number theory, algebraic geometry, and algebraic topology, and a proof was completed by Ngô Bao Châu in 2009. Ngô was promptly awarded a Fields medal for this work. (The proof ranked seventh on *Time* magazine's list of the top 10 scientific discoveries of the year, not bad for a result whose statement alone must have been out of reach for the vast majority of *Time*'s readers.)

Any actual statement or sketch of the fundamental lemma would require many pages to be comprehensible. An excellent attempt to provide a rough understanding can be found in a recent expository paper.[6] Although it is not practical here to attempt a conceptual picture of the Fundamental Lemma or the implications of its proof, this topic demonstrates that very difficult problems continue to be solved, while also providing fresh insights for further research. This accomplishment is strong evidence of the continuing vitality of the mathematical sciences.

It is interesting to note that, having grown up in Vietnam and studied in France, Ngô did his greatest work in the United States and is now a professor at the University of Chicago. This is an example of the way in which

---

[5] The proof of Fermat's Last Theorem was important and exciting for mathematicians because of the way the proof exposed deep connections between different areas of knowledge, not because people doubted the truth of the Theorem. By generalizing beyond the case of $n = 2$, the Langlands program could similarly reveal exciting and important connections and, with them, deep understanding.

[6] David Nadler, 2012, The geometric nature of the fundamental lemma. *Bulletin of the American Mathematical Society* 49(1):1-50.

the powerful and appealing culture of mathematics in the United States attracts some of the greatest scientists in the world to come and settle here.

PRIMES IN ARITHMETIC PROGRESSION

The prime numbers—whole numbers divisible only by themselves and 1—have been a focus of mathematical attention since the time of Euclid or before. This is partly because the primes are the building blocks of all numbers in a very precise sense; partly because the collection of primes displays a mixture of regularity and randomness that has been irresistibly tempting to curious mathematicians; and partly, in recent times, because of their amazing applications in computer science, especially cryptography.

Euclid's work contains a proof that there are infinitely many primes. One very old puzzle is whether there are infinitely many pairs of primes whose difference is only 2, such as 5 and 7, or 11 and 13. In spite of the ease with which this question is stated, it remains unsolved. Another old question, raised by Lagrange and Waring in 1770, concerns primes in arithmetic progression. One form of this question is easy to state: Is there a prime number p and some number q so that every one of the million numbers in the arithmetic progression

$$p, p + q, p + 2q, \ldots, p + 999,999q$$

is prime? Of course one could ask this same question with any number N in place of a million.

There was very little progress on this question until Ben Green and Terence Tao (who won a Fields medal based largely on his part in the work) proved that, indeed, for *any* N there are arithmetic progressions of N primes as above. In fact they proved much more; as with virtually every major mathematical advance, there is really a whole family of results, of which this is just an easy-to-state sample.

Why have we included this example to illustrate the vitality of the mathematical sciences? Because a key feature of the advance is the forging of a surprising link between prime numbers and two apparently unrelated fields of mathematics, harmonic analysis and ergodic theory. In this sense, the result of Green and Tao is typical of many great advances in mathematics, which seem to come from nowhere and combine apparently unrelated fields, opening new opportunities in the process.

Indeed, ergodic theory is generally considered a part of the mathematical theory of probability. The fact that it is useful in the study of prime numbers reflects the tension between treating primes as a completely deterministic phenomenon and recognizing that we can deal with them best by pretending that they are, in some ways, very random and thus best viewed through the lens of probability theory. Green and Tao made progress by showing that even a very disordered set, like the set of primes, can sometimes be decomposed into a highly structured part and a part that behaves in a highly random way.

For a detailed but still relatively comprehensible treatment of this development the reader may consult the 2005 paper of Bryna Kra.[7]

HIERARCHICAL MODELING

Hierarchical modeling (HM) is a set of techniques used in two related ways: to estimate characteristics of population distributions such as means and variances, often through combining information from different sources and to forecast individual characteristics taken from the population. As an intuitive example of how HM works, one can take the problem of ranking batters in baseball after a few games, based on the proportions of times at bat that batters succeeded in getting a hit. The proportion (or batting average) is partly reflective of the ability of the batter, but it also contains a great deal of randomness because only a few games have been played. It is well recognized that the initially very high (or very low) batting averages of many players will revert to the mean as the season progresses, but it is not always recognized that this is an almost inevitable process owing to the large component of randomness that is present initially.

Such situations are naturally modeled by an HM. One assumes that there is an unknown "true" batting average for each batter, based on skill level, and then observes a combination of this true average and measurement error. The measurement error is modeled (typically, in this case, by a simple Bernoulli model), and the true batting averages are also modeled as arising from a "population distribution" (the unknown distribution of true batting averages); it is this second-level modeling that leads to the name "hierarchical modeling." There are various possible ways to analyze the resulting model, but all can lead to interesting and surprising conclusions, such as a possible "cross-over effect," wherein a batter with a higher current average but fewer games played can be predicted to have less skill than a batter with a lower current average but more games played (because the random component of this latter batter's current average would be smaller). Of course, baseball aficionados would find such reversals to be reasonable.

While HM was introduced over 40 years ago, it became central to statistics and other sciences primarily over the last decade, as its computational demands have become tractable. In various contexts, hierarchical models have been referred to as random effects models, empirical Bayes

---

[7] Bryna Kra, 2005, The Green-Tao theorem on arithmetic progression in the primes: An ergodic point of view. *Bulletin* (New Series) *of the American Mathematical Society* 43(1):3-23.

models, multilevel models, random coefficient models, shrinkage methods, hidden Markov models, and many other terms. The surge in development and use of HM coincided with the surge in computational power, as well as an explosion of accompanying theoretical and algorithmic developments going under the name of Markov chain Monte Carlo (MCMC) methods. The following illustrations indicate how HM is being used in science and society today:

- *Climate and environmental studies often require the derivation of climate variables such as temperature or precipitation based on data from various sources.* For example, in paleoclimate reconstruction, a climate field needs to be recovered from different types of observations such as tree rings and pollen records from lake sediments or ice cores, as well as information about external forcings. (The three main forcings are volcanism, solar irradiance, and greenhouse gases.) Any model that relates tree ring widths to climate conditions is imperfect and thus contains uncertainty; other uncertainties are introduced through factors such as measurement errors and the changing number of trees as a function of time. The HM framework provides a set of probability models to link different observations to the climate process and to model uncertainties at different levels. The estimated latent climate variables together with their corresponding uncertainties can be assessed explicitly. Since the climate and environmental data are typically from multiple sources, it is rarely possible for a single model to feature all different types of data and reflect their intricate relationships. On the other hand, HM is particularly useful for modeling their complex structures and integrating all the information to produce a coherent solution for the unknown climate process.

- *Computational biologists use HM methods to analyze microarray data and to study patterns in genomic sequences of various species.* For example, one can assume a latent hidden Markov structure for a protein sequence and use observations from multiple species or multiple analogous copies from one species to find common conserved parts of the protein, which often correspond to functionally critical regions and can be informative for drug design. A similar structure can also be designed for control regions (called promoters) of multiple coregulated genes so as to discover binding sites of transcription factors. Algorithms such as BioProspector, MEME, AlignACE, and MDscan are all successful implementations of such models and have been widely used by biologists. A very exciting tool resulting from such models is GENSCAN, which is highly successful for ab initio prediction of complete gene structures in vertebrate, drosophila, and plant genomic sequences. Scientists also create HM structures to relate variances of the expressions of different genes so as to more accurately identify genes that behave (express) differently under different conditions—for example, in a cancerous fashion rather than a normal one.

- *An extension of the basic HM structure is the Bayesian network (BN), which has become an important machine learning tool in artificial intelligence.* A BN is a graphical approach to encode a probabilistic dependence model, in which each node in a graph represents a random variable that may or may not be observed and a directed link between two nodes indicates that they are dependent. Researchers find this structure very rich and efficient in learning relationships among many factors and making highly accurate predictions in complex situations. For example, scientists and engineers use BN for building spam-filtering tools, for information retrieval, image processing, biosurveillance, decision support systems, and so on.

- *Public health researchers, Census Bureau scientists, and geographers use HM for spatial analysis, such as for disease mapping in regions and for demographic estimates in small areas.* Actuaries refer to shrinkage estimation as credibility rate making in order to estimate the relative risks of different insured groups. Epidemiologists use these models for multiple comparisons and to help assess environmental risk. Economists and financial researchers improve population estimates via multilevel random regression coefficient models. The Food and Drug Administration uses these models to monitor the complexities of adverse drug reaction reports. The list goes on and on, and these applications will only grow as more data analysts reliably learn to use HMs, and as theoretical, algorithmic, and computational research continues to enlarge the domain of HM's applicability.

ALGORITHMS AND COMPLEXITY

Underlying much of engineering are algorithms that solve problems, often problems with deep and interesting mathematical structure. In recent years there have been significant improvements in our ability to solve such problems efficiently and to understand the limits of what is solvable.

Classical examples of deep algorithmic questions arise in all parts of the transportation industry, from airlines to package delivery. Algorithmic tools built on mathematical frameworks are used to design and update schedules and to plan routes with low demands on resources. Such algorithmic questions are not limited to transportation but are important in almost all large-scale business decisions made across most industries. Over the last few decades the mathematical science community made great strides in developing and improving algorithmic tools and making these available for wide use through companies like IBM—for instance, as embodied in its CPLEX optimization software package—or start-ups like Gurobi. The algorithms used for optimization all are based on deep mathematical ideas, although their efficiency has often been established only experimentally. Recent work initiated by Spielman and Teng on smoothed analysis[8] gives a new framework for understanding the efficiency of these methods, one that estimates performance probabilistically rather than focusing on the rarely seen worst-case outcomes.

Some algorithmic decisions need to be made instantaneously, without the benefit of information about trade-offs that is commonly part of tools for business optimization decisions. An environment where such optimizations are commonplace is scheduling of bits on the Internet. An example is the method used by Akamai, which serves almost all of the large Web sites. The company was based on a very theoretical algorithmic idea of how to best distribute content over the Internet.

Perhaps the best-known mathematical algorithm of great commercial interest is the RSA encryption scheme. The RSA system dates back to the mid-1970s, and its developers—Ron Rivest, Adi Shamir, and Leonard Adleman—won the 2002 Turing award for their discovery. Encryption is now the basis of all Internet commerce, used not only for encrypting messages but also for all forms of sensitive communications including authentication, digital signatures, digital cash, and electronic voting, to name just a few. Our understanding of how to make such applications possible has grown tremendously in recent years. New encryption schemes have been developed, providing encryptions that are more secure or simpler to use, and we better understand the level of security provided by encryption schemes in increasingly complex environments with many new forms of attacks. Maybe the most fundamental of these developments is the discovery of what is known as fully homomorphic encryption, which allows one to compute with the encrypted information without decrypting the messages or learning their content.

Coding of information allows us to reliably transmit information across a noisy channel, later correcting the errors at the receiver, and to efficiently store and retrieve information. Such error-correcting codes are the basis of much of the digital communication technology underlying cell phones, videos, CDs, and so on. Shannon's classical work in this area used probability and statistics to bound the amount of information that can be

---

[8] Daniel A. Spielman and Shang-Hua Teng, 2004, Smoothed analysis of algorithms: Why the simplex algorithm usually takes polynomial time. *Journal of the ACM* 51(3):385-463.

transmitted in a noisy channel; this bound is now known as the Shannon capacity of the channel. Recent work by Spielman and others in the computer science theory community[9] has developed new codes that can be encoded and decoded in linear time and reach the Shannon capacity even against worst-case noise models. Coding has also had an important impact on many seemingly unrelated fields, such as understanding the limits of what is efficiently computable. Coding can be seen as the key tool in the development of a useful new type of proof system called probabilistically checkable proofs. These are proofs whose correctness can be verified only probabilistically; for example, a randomized algorithm that only inspects a few characters from the proof might show it to be correct with, say, 99 percent certainty. In a stunning development in complexity theory, it has been shown that every correct proof can be converted into a probabilistically checkable proof with only a polynomial increase in the size of the proof, effectively providing for a highly redundant coding of the proof. The theory of probabilistically checkable proofs is a key component in allowing us to prove the computational hardness of finding approximately optimal answers to classical optimization problems. To see the connection, observe that finding the most convincing probabilistically checkable proof is an approximation problem that is hard: true statements have probabilistically checkable proofs, and any probabilistic proof aiming to prove a false statement must fail with high probability.

INVERSE PROBLEMS: VISIBILITY AND INVISIBILITY

Inverse problems are those for which one creates an understanding of the internal structure of a system from external observations.[10] The system itself is hidden, a black box that cannot be probed directly: Examples are a patient undergoing a medical imaging procedure, a nontransparent industrial object whose internal structural integrity is being examined, or a structure beneath Earth's surface, such as a petroleum reservoir. One knows a lot about the physics of the system, but much about the structure and parameters needs to be reconstructed. Based on the way external signals are affected by passing through the black-box system, one needs to recover the unknown parameters of the system. Such problems lie at the heart of contemporary scientific inquiry and technological development. Applications include a vast variety of imaging techniques for purposes such as the early detection of cancer and pulmonary edema, locating oil and mineral

---

[9] See, for example, Dan Spielman, 1996, Linear-time encodable and decodable error-correcting codes. *IEEE Transactions on Information Theory* 42(6):1723-1731.

[10] The committee thanks Gunther Uhlmann of the University of Washington for contributing material for this section.

deposits in Earth's crust, creating astrophysical images from telescope data, finding cracks and interfaces within materials, optimizing shapes, identifying models in growth processes, and modeling in the life sciences.

A typical inverse problem calls for determining the coefficients of a partial differential equation given some information about the solutions to the equation. Research in this area draws on a diverse array of mathematics, including complex analysis, differential geometry, harmonic analysis, integral geometry, numerical analysis, optimization, partial differential equations, and probability, and it builds strong linkages between applications and deep areas of mathematics.

An archetypal example of an inverse boundary problem for an elliptic equation is the by-now-classical Calderón problem, also called electrical impedance tomography (EIT). Calderón proposed the problem in the mathematical literature in 1980. In EIT one attempts to determine the electrical conductivity of a medium by taking measurements of voltage and current at the boundary of the medium. The information is encoded in the Dirichlet to Neumann (DN) map associated with the conductivity equation. EIT arises in several applications, including geophysical prospecting and medical imaging. In the last 25 years or so there has been remarkable progress on this problem. In the last few years this includes addressing the two-dimensional problem, the case of partial data, and also the discrete problem.

New inverse problems arise all the time because there are so many applications. For example, in medical imaging, there has been considerable interest in recent years in hybrid methods, also called multiwave methods, which combine a high-resolution modality with a high-contrast one. For example, in breast imaging ultrasound provides a high (submillimeter) resolution, but it suffers from low contrast, whereas many tumors absorb much more energy from electromagnetic waves than do healthy cells, so that imaging with electromagnetic waves provides very high contrast.

Research on Calderón's problem led to new insight on the problem of making objects invisible to detection by electromagnetic waves, sound waves, and other types of waves. Invisibility has been a subject of human fascination for millennia, from the Greek legend of Perseus and Medusa to the more recent Harry Potter stories. Since 2003 there has been a wave of serious theoretical proposals in the mathematics and physics literature about cloaking devices—structures that would not only render an object invisible but also undetectable to electromagnetic waves. The particular route to cloaking that has received the most attention is that of transformation optics, the design of optical devices with customized effects on wave propagation. Transformation optics was developed first for the case of electrostatics—that is, for Calderón's problem. Transformation optics takes advantage of the transformation rules for the material properties of optics: the index of refraction for scalar optics and acoustics, governed by the Helmholtz equation, and the electric permittivity and magnetic permeability for vector optics, as described by Maxwell's equations. The index of refraction of these materials is such that light rays go around a region and emerge on the other side as if they had passed through empty space. Advances in metamaterials (material not found in nature) have made possible the realization for microwave frequencies of the blueprints for cloaking found theoretically using transformation optics. There has been a concerted effort in recent years in constructing these materials for a broader range of frequencies, including visible light. Metamaterials were identified by *Science* as one of the top 10 insights of the last decade.[11]

THE INTERPLAY OF GEOMETRY AND THEORETICAL PHYSICS

There is a long history of interactions of mathematics, and in particular geometry, with theoretical physics. At one point in the mid-nineteenth century the fields were one and the same. For example, Dirichlet's argument for the existence of harmonic functions on the disk with given boundary values on the circle was an appeal to physical intuition about electrostatics. One model for the interaction is seen in both the final formulations of quantum mechanics and of general relativity. By the late 1920s, after many false starts and partial formulations, quantum mechanics was finally formulated in terms of Hilbert spaces and operators acting on these spaces. These mathematical objects had been introduced by Hilbert in the 1880s for purely mathematical reasons having nothing to do with quantum mechanics, which had not even been conceived of then. It is interesting to note, though, that Hilbert called the decomposition of his operators the "spectral decomposition" because it reminded him of the spectrum of various atoms, something that was mysterious at the time and finally explained by quantum mechanics. Einstein struggled for many years to formulate general relativity without finding the appropriate mathematical context. Finally, he learned of Riemann's work on what is now called Riemannian geometry. This was exactly the formalism he was looking for, and soon after learning about this theory he had formulated general relativity in terms of a four-dimensional space-time continuum with an indefinite Riemannian metric which is positive in the time directions and negative in the spatial directions. In each of these cases the mathematics in question had been developed for internal mathematical reasons and its existence predated the need for it in physics. It was there ready to use when physics was ready for it. Interactions such as these led the physicist

---

[11] This paragraph draws on Allen Greenleaf, Yaroslav Kurylev, Matti Lassas, and Gunther Uhlmann, 2009, Cloaking devices, electromagnetic wormholes, and transformation optics. *SIAM Review* 51 (March):3.

Eugene Wigner to wonder what accounts for the unreasonable effectiveness of mathematics in physics.

A more recent version of the same basic pattern is the Yang-Mills theory. Here again the physicists were struggling to develop a mathematical framework to handle the physical concepts they were developing, when in fact the mathematical framework, which in mathematics is known as connections on principal bundles and curvature, had already been introduced for mathematical reasons. Much of the recent history of quantum field theory has turned this model of interaction on its head. When quantum field theory was introduced in the 1940s and 1950s there was no appropriate mathematical context. Nevertheless, physicists were able to develop the art of dealing with these objects, at least in special cases. This line of reasoning, using as a central feature the Yang-Mills theory, led to the standard model of theoretical physics, which makes predictions that have been checked by experiment to enormous precision. Nevertheless, there was not then and still is not today a rigorous mathematical context for these computations. The situation became even worse with the advent of string theory, where the appropriate mathematical formulation seems even more remote. But the fact that the mathematical context for these theories did not exist and has not yet been developed is only part of the way that the current interactions between mathematics and physics differ from previous ones. As physicists develop and explore these theories, for which no rigorous mathematical formulation is known, they have increasingly used ever more sophisticated geometric and topological structures in their theories. As physicists explore these theories they come across mathematical questions and statements about the underlying geometric and topological objects in terms of which the theories are defined. Some of these statements are well-known mathematical results, but many turn out to be completely new types of mathematical statements.

These statements, conjectures, and questions have been one of the main forces driving geometry and topology for the last 20 to 25 years. Some of them have been successfully verified mathematically; some have not been proved but the mathematical evidence for them is overwhelming; and some are completely mysterious mathematically. One of the first results along these lines gave a formula for the number of lines of each degree in a general hypersurface in complex projective four-dimensional space given by a homogeneous degree-5 polynomial. Physics arguments produced a general formula for the number of such lines where the formula came from a completely different area of mathematics (power series solutions to certain ordinary differential equations). Before the input from physics, mathematicians had computed the answer for degrees 1 through 5 but had no conjecture for the general answer. The physics arguments provided the general formula, and this was later verified by direct mathematical argument.

These direct mathematical arguments gave no understanding of the original physics insight that connected the formula with solutions to an ordinary differential equation. Indeed, finding such a connection is one of the central problems today in geometry. Many mathematicians work on aspects of this problem, and there are partial hints but no complete understanding, even conjecturally. This statement characterizes much of the current input of physics into mathematics. It seems clear that the physicists are on the track of a deeper level of mathematical understanding that goes far beyond our current knowledge, but we have only the smallest hints of what that might be. Understanding this phenomenon is a central problem both in high-energy theoretical physics and in geometry and topology.

Nowadays, sophisticated mathematics is essential for stating many of the laws of physics. As mentioned, the formulation of the standard model of particle physics involves "gauge theories," or fiber bundles. These have a very rich topology. These topological structures are described by Chern-Simons theories, Index theory, and K-theory. These tools are also useful for condensed matter systems. They characterize topological phases of matter, which could offer an avenue for quantum computing.[12] Here the q-bits are encoded into the subtle topology of the fiber bundle, described by Chern-Simons theory. Recently, K-theory has been applied to the classification of topological insulators,[13] another active area of condensed matter physics.

String theory and mathematics are very closely connected, and research in these areas often straddles physics and mathematics. One recent development, the gauge gravity duality, or AdS/CFT, has connected general relativity with quantum field theories, the theories we use for particle physics.[14] The gravity theory lives in hyperbolic space. Thus, many developments in hyperbolic geometry, and black holes, could be used to describe certain strongly interacting systems of particles. Thinking along these lines has connected a certain long-distance limit of gravity equations to the equations of hydrodynamics. One considers a particular black-hole, or black-brane, solution of Einstein's equations with a negative cosmological constant. These black holes have long-distance excitations representing small fluctuations of the geometry. The fluctuations are damped since they end up being swallowed by the black hole. According to AdS/CFT, this system is described by a thermal system on the boundary, a thermal fluid of quantum interacting particles. In this formulation, the long-distance perturbations are described by hydrodynamics, namely by the Navier-Stokes equation

[12] A. Yu. Kitaev, 2003, Fault-tolerant quantum computation by anyons. *Annals of Physics* 303:2-30

[13] A.P. Schnyder, S. Ryu, A. Furusaki, and A.W.W. Ludwig, 2008, Classification of topological insulators and superconductors in three spatial dimensions. *Physical Review Letters B* 78:195125.

[14] J. Maldacena, 2005, The illusion of gravity. *Scientific American* October 24.

or its relativistic analog. The viscosity term in this equation produces the damping of excitations, and it is connected with the waves falling into the black hole. Computing the viscosity in the particle theory from first principles is very difficult. However, it is very simple from the point of view of Einstein's equations, because it is given by a purely geometric quantity: the area of the black hole horizon. This has been used to qualitatively model strongly interacting systems of quantum particles. These range from the quark-gluon fluids that are produced by heavy ion collisions (at the Relativistic Heavy Ion Collider at Brookhaven National Laboratory or at the Large Hadron Collider at Geneva) to high-temperature superconductors in condensed matter physics.

Many examples of AdS/CFT involve additional structures, such as supersymmetry. In these cases the geometry obeys special constraints, giving rise to Sasaki-Einstein spaces, which are closely related to Calabi-Yau spaces. This is merely an example of a more general trend developing connections between matrix models, algebraic curves, and supersymmetric quantum field theory.

A recent development of the past decade has been the discovery of integrability in $N = 4$ super-Yang-Mills. This four-dimensional quantum field theory is the most symmetric quantum field theory. The study of this highly symmetrical example is very useful since it will probably enable us to find some underlying structures common to all quantum gauge theories. Integrability implies the existence of an infinite-dimensional set of symmetries in the limit of a large number of colors. In this regime the particles of the theory, or gluons, form a sort of necklace. The symmetry acts on these states and allows us to compute their energies exactly as a function of the coupling. The deep underlying mathematical structures are only starting to be understood. Integrability in so-called $(1 + 1)$-dimensional systems has led to the development of quantum groups and other interesting mathematics. The way integrability appears here is somewhat different, and it is quite likely that it will lead to new mathematics. A closely related area is the computation of scattering amplitudes in this theory. A direct approach using standard methods, such as Feynman diagrams, quickly becomes very complicated. On the other hand, there are new methods showing that the actual answers are extremely simple and have a rich structure that is associated with the mathematics of Grassmanians. This has led to another fruitful collaboration.[15]

The connection between theoretical physics and mathematics is growing ever stronger, and it is supported by the emergence of interdisciplinary centers, such as the Simons Center for Geometry and Physics at Stony

---

[15] See, for example, A.B. Goncharov, M. Spradlin, C. Vergu, and A. Volovich, 2010, Classical polylogarithms for amplitudes and Wilson loops. *Physical Review Letters* 105:151605.

Brook, and math/physics initiatives at various universities, such as Duke University, the University of Pennsylvania, and the University of California at Santa Barbara.

NEW FRONTIERS IN STATISTICAL INFERENCE

We live in a new age for statistical inference, where technologies now produce high-dimensional data sets, often with huge numbers of measurements on each of a comparatively small number of experimental units. Examples include gene expression microarrays monitoring the expression levels of tens of thousands of genes at the same time and functional magnetic resonance imaging machines monitoring the blood flow in various parts of the brain. The breathtaking increases in data-acquisition capabilities are such that millions of parallel data sets are routinely produced, each with its own estimation or testing problem. This era of scientific mass production calls for novel developments in statistical inference, and it has inspired a tremendous burst in statistical methodology. More importantly, the data flood completely transforms the set of questions that needs to be answered, and the field of statistics has, accordingly, changed profoundly in the last 15 years. The shift is so strong that the subjects of contemporary research now have very little to do with general topics of discussion from the early 1990s.

High dimensionality refers to an estimation or testing problem in which the number of parameters about which we seek inference is about the same as, or much larger than, the number of observations or samples we have at our disposal. Such problems are everywhere. In medical research, we may be interested in determining which genes might be associated with prostate cancer. A typical study may record expression levels of thousands of genes for perhaps 100 men, of whom only half have prostate cancer and half serve as a control set. Here, one has to test thousands of hypotheses simultaneously in order to discover a small set of genes that could be investigated for a causal link to cancer development. Another example is genomewide association studies where the goal is to test whether a variant in the genome is associated with a particular phenotype. Here the subjects in a study typically number in the tens of thousands and the number of hypotheses may be anywhere from 500,000 to 2.5 million. If we are interested in a number of phenotypes, the number of hypotheses can easily rise to the billions and trillions, not at all what the early literature on multiple testing had in mind.

In response, the statistical community has developed groundbreaking techniques such as the false discovery rate (FDR) of Benjamini and Hochberg, which proposes a new paradigm for multiple comparisons and has had a tremendous impact not only on statistical science but also in the medical sciences and beyond.[16] In a nutshell, the FDR procedure controls the expected ratio between the number of false rejections and the total number of rejections. Returning to our example above, this allows the statistician to return a list of genes to the medical researcher assuring her that she should expect at most a known fraction of these genes, say 10 percent, to be "false discoveries." This new paradigm has been extremely successful, for it enjoys increased power (the ability of making true discoveries) while simultaneously safeguarding against false discoveries. The FDR methodology assumes that the hypotheses being tested are statistically independent and that the data distribution under the null hypothesis is known. These assumptions may not always be valid in practice, and much of statistical research is concerned with extending statistical methodology to these challenging setups. In this direction, recent years have seen a resurgence of empirical Bayes techniques, made possible by the onslaught of data and providing a powerful framework and new methodologies to deal with some of these issues.

Estimation problems are also routinely high-dimensional. In a genetic association study, n subjects are sampled and one or more quantitative traits, such as cholesterol level, are recorded. Each subject is also measured at p locations on the chromosomes. For instance, one may record a value (0, 1, or 2) indicating the number of copies of the less-common allele observed. To find genes exhibiting a detectable association with the trait, one can cast the problem as a high-dimensional regression problem. That is to say, one seeks to express the response of interest (cholesterol level) as a linear combination of the measured genetic covariates; those covariates with significant coefficients are linked with the trait.

The issue is that the number n of samples (equations) is in the thousands while the number p of covariates is in the hundreds of thousands. Hence, we have far fewer equations than unknowns, so what shall we do? This is a burning issue because such underdetermined systems arise everywhere in science and engineering. In magnetic resonance imaging, for example, one would like to infer a large number of pixels from just a small number of linear measurements. In many problems, however, the solution is assumed to be sparse. In the example above, it is known that only a small number of genes can potentially be associated with a trait. In medical imaging, the image we wish to form typically has a concise description in a carefully chosen representation.

In recent years, statisticians and applied mathematicians have developed a flurry of highly practical methods for such sparse regression problems. Most of these methods rely on convex optimization, a field that has

---

[16] As of January 15, 2012, Google Scholar reported 12,861 scientific papers citing the original article of Benjamini and Hochberg.

registered spectacular advances in the last 15 years. By now there is a large literature explaining (1) when one can expect to solve a large underdetermined system by L1 minimization and when it is not possible and (2) when accurate statistical estimation is possible. Beyond the tremendous technical achievements, which have implications for nearly every field of science and technology, this modern line of research suggests a complete revision of those metrics with which the accuracy of statistical estimates is evaluated. Whereas classical asymptotic approaches study the size of errors when the number of parameters is fixed and the sample size goes to infinity, modern asymptotics is concerned with situations when the number of both parameters p and observations n tends to infinity but perhaps in a fixed ratio, or with p growing at most polynomially in n. Further, the very concept of errors has to be rethought. The question is not so much whether asymptotic normality holds but whether the right variables have been selected. In a sea of mostly irrelevantvariables, collected because data are now so readily acquired in many contexts. What is the minimum signal-to-noise ratio needed to guarantee that the variables of true importance can be identified?

Accurate estimation from high-dimensional data is not possible unless one assumes some structure such as sparsity, discussed above. Statisticians have studied other crucial structures that permit accurate estimation, again, from what appear to be incomplete data. These include the estimation of low-rank matrices, as in the famous Netflix problem, where the goal is to predict preferences for movies a user has not yet seen; the estimation of large covariance matrices or graphical models under the assumption that the graph of partial correlations is sparse; or the resolution of X-ray diffraction images from magnitude measurements only. The latter is of paramount importance in a number of applications where a detector can collect only the intensity of an optical wave but not its phase. In short, modern statistics is at the forefront of contemporary science since it is clear that progress will increasingly rely on statistical and computational tools to extract information from massive data sets.

Statistics has clearly taken on a pivotal role in the era of massive data production. In the area of multiple comparisons, novel methodologies have been widely embraced by applied communities. Ignoring statistical issues is a dangerous proposition readily leading to flawed inference, flawed scientific discoveries, and nonreproducible research. In the field of modern regression, methods and findings have inspired a great number of communities, suggesting new modeling tools and new ways to think about information retrieval. The field is still in its infancy, and much work remains to be done. To give one idea of a topic that is likely to preoccupy statisticians in the next decade, one could mention the problem of providing correct inference after selection. Conventional statistical inference requires that a model be known before the data are analyzed. Current practice, however, typically selects a model after data analysis. Standard statistical tests and confidence intervals applied to the selected parameters are in general completely erroneous. Statistical methodology providing correct inference after massive data snooping is urgently needed.

ECONOMICS AND BUSINESS: MECHANISM DESIGN

Mechanism design is a subject with a long and distinguished history. One such example is designing games that create incentives to produce a certain desired outcome (such as revenue or social welfare maximization). Recent developments highlight the need to add computational considerations to classic mechanism-design problems.

Perhaps the most important example is the sale of advertising space on the Internet, which is the primary source of revenue for many providers of online services. According to a recent report, online advertising spending continues to grow at double-digit rates, with $25.8 billion having been spent in online advertising in the United States in 2010.[17] The success of online advertising is due, in large part, to providers' ability to tailor advertisements to the interests of individual users as inferred from their search behavior. However, each search query generates a new set of advertising spaces to be sold, each with its own properties determining the applicability of different advertisements, and these ads must be placed almost instantaneously. This situation complicates the process of selling space to potential advertisers.

There has been significant progress on computationally feasible mechanism design on many fronts. Three highlights of this research so far are the following:

- Understanding the computational difficulty of finding Nash equilibria. Daskalakis, Goldberg, and Papadimitriou won the Game Theory Society's Prize in Game Theory and Computer Science in 2008 for this work.
- Quantifying the loss of efficiency of equilibria in games that do not perfectly implement a desired outcome, which is referred to as the price of anarchy. While initial success in this area has been in the area of games such as load balancing and routing, recent work in this direction that is relevant to online auctions also has great potential.
- Enabling computationally feasible mechanism design by developing techniques that approximately implement a desired outcome.

---

[17] Data from emarketer.com. Available at http://www.emarketer.com/PressRelease.aspx?R=1008096. Accessed March 14, 2012.

MATHEMATICAL SCIENCES AND MEDICINE

The mathematical sciences contribute to medicine in a great many ways, including algorithms for medical imaging, computational methods related to drug discovery, models of tumor growth and angiogenesis, health informatics, comparative effectiveness research, epidemic modeling, and analyses to guide decision making under uncertainty. With the increasing availability of genomic sequencing and the transition toward more widespread use of electronic health record systems, we expect to see more progress toward medical interventions that are tailored to individual patients. Statisticians will be deeply involved in developing those capabilities.

To illustrate just one way in which these endeavors interact, consider some mathematical science challenges connected to the diagnosis and planning of surgery for cardiac patients. One of the grand challenges of computational medicine is how to construct an individualized model of the heart's biology, mechanics, and electrical activity based on a series of measurements taken over time. Such models can then be used for diagnosis or surgical planning to lead to better patient outcomes. Two basic mathematical tasks are fundamental to this challenge. Both are much-studied problems in applied mathematics, but they need to be carefully adapted to the task at hand.

The first of these tasks is to extract cardiac motion from a time-varying sequence of three-dimensional computerized tomography (CT) or magnetic resonance imaging (MRI) patient images. This is achieved by solving the so-called deformable image registration problem, a problem that comes up over and over in medical imaging. To solve this problem—to effectively align images in which the principal subject may have moved—one needs to minimize a chosen "distance" between the image intensity functions of images taken at different times. Unfortunately, this problem is ill-posed: there are many different maps that minimize the distance between the two images, most of which are not useful for our purposes. To tease out the appropriate mapping, one must choose a "penalty function" for the amount of stretching that is required to bring the successive images into approximate alignment. Finding the right penalty function is a very subtle task that relies on concepts and tools from a branch of core mathematics called differential geometry. Once a good penalty function has been applied, the work also requires efficient computational algorithms for the large-scale calculations.

The second mathematical task is to employ the extracted cardiac motion as observational data that drive the solution of an inverse problem. In this case, the inverse problem is to infer the parameters for the bioelectromechanical properties of the cardiac model based on the motion observed externally through the imaging. The cardiac biophysical model draws on another area of mathematics—partial differential equations—and must bring together multiple physics components: elasticity of the ventricular wall, electrophysiology, and active contraction of the myocardial fibers.

The full-blown setting of this problem is analogous to a "blind deconvolution" problem, in the sense that neither the model nor the source is fully known. As such, this presents enormous difficulty for the inversion solvers; as in the image registration case, it requires careful formulation and regularization, as well as large-scale computational solvers that are tolerant of ill-conditioning. Recent research[18] is following a hybrid approach that interweaves the solution of the image registration and model determination problems.

COMPRESSED SENSING

The story of compressed sensing is an example of the power of the mathematical sciences and of their dynamic relationship with science and engineering. As is often the case, the development of novel mathematics can be inspired by an important scientific or engineering question. Then, mathematical scientists develop abstractions and quantitative models to solve the original problem, but the conversion into a more abstract setting can also supply insight to other applications that share a common mathematical structure. In other words, there is no need to reinvent the wheel for each instantiation of the problem.

Compressed sensing was motivated by a great question in MRI, a medical imaging technique used in radiology to visualize detailed internal structures. MRI is a wonderful tool with several advantages over other medical imaging techniques such as CT or X-rays. However, it is also an inherently slow data-acquisition process. This means that it is not feasible to acquire high-quality scans in a reasonable amount of time, or to acquire dynamic images (videos) at a decent resolution. In pediatrics for instance, the impact of MRI on children's health is limited because, among other things, children cannot remain still or hold their breath for long periods of time, so that it is impossible to achieve high-resolution scans. This could be overcome by, for example, using anesthesia that is strong enough to stop respiration for several minutes, but clearly such procedures are dangerous.

Faster imaging can be achieved by reducing the number of data points that need to be collected. But common wisdom in the field of biomedical imaging maintained that skipping sample points would result in information loss. A few years ago, however, a group of researchers turned signal processing upside down by showing that high-resolution imaging was pos-

[18] H. Sundar, C. Davatzikos, and G. Biros, 2009, Biomechanically-constrained 4D estimation of myocardial motion. *Medical Image Computing and Computer-Assisted Intervention* (MICCAI):257-265.

sible from just a few samples. In fact, they could recover high-resolution pictures even when an MRI is not given enough time to complete a scan. To quote from *Wired*, "That was the beginning of compressed sensing, or CS, the paradigm-busting field in mathematics that's reshaping the way people work with large data sets."[19]

Despite being only a few years old, compressed sensing algorithms are already in use in some form in several hospitals in the country. For example, compressed sensing has been used clinically for over 2 years at Lucile Packard Children's Hospital at Stanford. This new method produces sharp images from brief scans. The potential for this method is such that both General Electric and Phillips Corporation have medical imaging products in the pipeline that will incorporate compressed sensing.

However, what research into compressed sensing discovered is not just a faster way of getting MR images. It revealed a protocol for acquiring information, all kinds of information, in the most efficient way. This research addresses a colossal paradox in contemporary science, in that many protocols acquire massive amounts of data and then discard much of it, without much or any loss of information, through a subsequent compression stage, which is usually necessary for storage, transmission, or processing purposes. Digital cameras, for example, collect huge amounts of information and then compress the images so that they fit on a memory card or can be sent over a network. But this is a gigantic waste. Why bother collecting megabytes of data when we know very well that we will throw away 95 percent of it? Is it possible to acquire a signal in already compressed form? That is, Can we directly measure the part of the signal that carries significant information and not the part of the signal that will end up being thrown away? The surprise is that mathematical scientists provide an affirmative answer. It was unexpected and counterintuitive, because common sense says that a good look at the full signal is necessary in order to decide which bits one should keep or measure and which bits can be ignored or discarded. This view, although intuitive, is wrong. A very rich mathematical theory has emerged showing when such compressed acquisition protocols are expected to work.

This mathematical discovery is already changing the way engineers think about signal acquisition in areas ranging from analog-to-digital conversion, to digital optics, and seismology. In communication and electronic intelligence, for instance, analog-to-digital conversion is key to transducing information from complex radiofrequency environments into the digital domain for analysis and exploitation. In particular, adversarial communications can hop from frequency to frequency. When the frequency range is large, no analog-to-digital converter (ADC) is fast enough to scan the

---

[19] Jordan Ellenberg, 2010, Fill in the blanks: Using math to turn lo-res datasets into hi-res samples. *Wired*, March.

full range, and surveys of high-speed ADC technologies show that they are advancing at a very slow rate. However, compressed sensing ideas show that such signals can be acquired at a much lower rate, and this has led to the development of novel ADC architectures aiming at the reliable acquisition of signals that are in principle far outside the range of current data converters. In the area of digital optics, several systems have been designed. Guided by compressed sensing research, engineers have more design freedom in three dimensions: (1) they can consider high-resolution imaging with far fewer sensors than were once thought necessary, dramatically reducing the cost of such devices; (2) they can consider designs that speed up signal acquisition time in microscopy by orders of magnitude, opening up new applications; and (3) they can sense the environment with greatly reduced power consumption, extending sensor life. Remarkably, a significant fraction of this work takes place in industry, and a number of companies are already engineering faster, cheaper, and more-efficient sensors based on these recently developed mathematical ideas.

Not only is compressed sensing one of the most applicable theories coming out of the mathematical sciences in the last decade, but it is also very sophisticated mathematically. Compressed sensing uses techniques of probability theory, combinatorics, geometry, harmonic analysis, and optimization to shed new light on fundamental questions in approximation theory: How many measurements are needed to recover an object of interest? How is recovery possible from a minimal number of measurements? Are there tractable algorithms to retrieve information from condensed measurements? Compressed sensing research involves the development of mathematical theories, the development of numerical algorithms and computational tools, and the implementation of these ideas into novel hardware. Thus, progress in the field involves a broad spectrum of scientists and engineers, and core and applied mathematicians, statisticians, computer scientists, circuit designers, optical engineers, radiologists, and others regularly gather to attend scientific conferences together. This produces a healthy cycle in which theoretical ideas find new applications and where applications renew theoretical mathematical research by offering new problems and suggesting new directions.

3

Connections Between the Mathematical Sciences and Other Fields

INTRODUCTION

In addition to ascertaining that the internal vitality of the mathematical sciences is excellent, as illustrated in Chapter 2, the current study found a striking expansion in the impact of the mathematical sciences on other fields, as well as an expansion in the number of mathematical sciences subfields that are being applied to challenges outside of the discipline. This expansion has been ongoing for decades, but it has accelerated greatly over the past 10-20 years. Some of these links develop naturally, because so much of science and engineering now builds on computation and simulation for which the mathematical sciences are the natural language. In addition, data-collection capabilities have expanded enormously and continue to do so, and the mathematical sciences are innately involved in distilling knowledge from all those data. However, mechanisms to facilitate linkages between mathematical scientists and researchers in other disciplines must be improved.

The impacts of mathematical science research can spread very rapidly in some cases, because a new insight can quickly be embodied in software without the extensive translation steps that exist between, say, basic research in chemistry and the use of an approved pharmaceutical. When mathematical sciences research produces a new way to compress or analyze data, value financial products, process a signal from a medical device or military system, or solve the equations behind an engineering simulation, the benefit can be realized quickly. For that reason, even government agencies or industrial sectors that seem disconnected from the mathematical sciences have a vested interest in the maintance of a strong mathematical sciences enterprise for our nation. And because that enterprise must be healthy in order to contribute to the supply of well-trained individuals in science, technology, engineering, and mathematical (STEM) fields, it is clear that everyone should care about the vitality of the mathematical sciences.

This chapter discusses how increasing interaction with other fields has broadened the definition of the mathematical sciences. It then documents the importance of the mathematical sciences to a multiplicity of fields. In many cases, it is possible to illustrate this importance by looking at major studies by the disciplines themselves, which often list problems with a large mathematical sciences component as being among their highest priorities. Extensive examples of this are given in Appendix D.

BROADENING THE DEFINITION OF THE MATHEMATICAL SCIENCES

Over the past decade or more, there has been a rapid increase in the number of ways the mathematical sciences are used and the types of mathematical ideas being applied. Because many of these growth areas are fostered by the explosion in capabilities for simulation, computation, and data analysis (itself driven by orders-of-magnitude increases in data collection), the related research and its practitioners are often assumed to fall within the umbrella of computer science. But in fact people with varied backgrounds contribute to this work. The process of simulation-based science and engineering is inherently very mathematical, demanding advances in mathematical structures that enable modeling; in algorithm development; in fundamental questions of computing; and in model validation, uncertainty quantification, analysis, and optimization. Advances in these areas are essential as computational scientists and engineers tackle greater complexity and exploit advanced computing. These mathematical science aspects demand considerable intellectual depth and are inherently interesting for the mathematical sciences.

At present, much of the work in these growth areas—for example, bioinformatics, Web-based companies, financial engineering, data analytics, computational science, and engineering—is handled primarily by people who would not necessarily be labeled "mathematical scientists." But the mathematical science content of such work, even if it is not research, is considerable, and therefore it is critical for the mathematical sciences community to play a role, through education, research, and collaboration. People with mathematical science backgrounds per se can bring different perspectives that complement those of computer scientists and others, and the combination of talents can be very powerful.

There is no precise definition of "the mathematical sciences." The following definition was used in the 1990 report commonly known as the David II report after the authoring committee's chair, Edward E. David:

> The discipline known as the mathematical sciences encompasses core (or pure) and applied mathematics, plus statistics and operations research, and extends to highly mathematical areas of other fields such as theoretical computer science. The theoretical branches of many other fields—for instance, biology, ecology, engineering, economics—merge seamlessly with the mathematical sciences.[1]

The 1998 Odom report implicitly used a similar definition, as embodied in Figure 3-1, adapted from that report.

Figure 3-1 captures an important characteristic of the mathematical sciences—namely, that they overlap with many other disciplines of science, engineering, and medicine, and, increasingly, with areas of business such as finance and marketing. Where the small ellipses overlap with the main ellipse (representing the mathematical sciences), one should envision a mutual entwining and meshing, where fields overlap and where research and people might straddle two or more disciplines. Some people who are clearly affiliated with the mathematical sciences may have extensive interactions and deep familiarity with one or more of these overlapping disciplines. And some people in those other disciplines may be completely comfortable in mathematical or statistical settings, as will be discussed further. These interfaces are not clean lines but instead are regions where the disciplines blend. A large and growing fraction of modern science and engineering is "mathematical" to a significant degree, and any dividing line separating the more central and the interfacial realms of the mathematical sciences is sure to be arbitrary. It is easy to point to work in theoretical physics or theoretical computer science that is indistinguishable from research done by mathematicians, and similar overlap occurs with theoretical ecology, mathematical biology, bioinformatics, and an increasing number of fields. This is not a new phenomenon—for example, people with doctorates in mathematics, such as Herbert Hauptman, John Pople, John Nash, and Walter Gilbert, have won Nobel prizes in chemistry or economics—but it is becoming more widespread as more fields become amenable to mathematical representations. This explosion of opportunities means that much of twenty-first century research is going to be built on a mathematical science foundation, and that foundation must continue to evolve and expand.

---

[1] NRC, 1990, *Renewing U.S. Mathematics: A Plan for the 1990s.* National Academy Press, Washington, D.C.

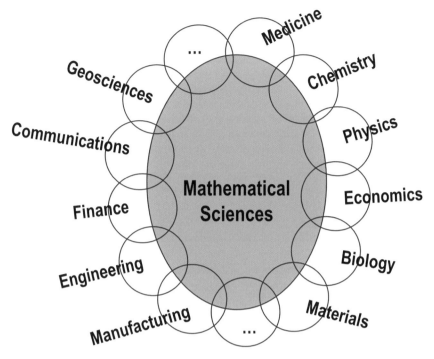

FIGURE 3-1 The mathematical sciences and their interfaces. SOURCE: Adapted from National Science Foundation, 1998, *Report of the Senior Assessment Panel for the International Assessment of the U.S. Mathematical Sciences*, NSF, Arlington, Va.

Note that the central ellipse in Figure 3-1 is not subdivided. The committee members—like many others who have examined the mathematical sciences—believe that it is important to consider the mathematical sciences as a unified whole. Distinctions between "core" and "applied" mathematics increasingly appear artificial; in particular, it is difficult today to find an area of mathematics that does not have relevance to applications. It is true that some mathematical scientists primarily prove theorems, while others primarily create and solve models, and professional reward systems need to take that into account. But any given individual might move between these modes of research, and many areas of specialization can and do include both kinds of work. Overall, the array of mathematical sciences share a commonality of experience and thought processes, and there is a long history of insights from one area becoming useful in another.

Thus, the committee concurs with the following statement made in the 2010 *International Review of Mathematical Sciences* (Section 3.1):

A long-standing practice has been to divide the mathematical sciences into categories that are, by implication, close to disjoint. Two of the most common distinctions are drawn between "pure" and "applied" mathematics, and between "mathematics" and "statistics." These and other categories can be useful to convey real differences in style, culture and methodology, but in the Panel's view, they have produced an increasingly negative effect when the mathematical sciences are considered in the overall context of science and engineering, by stressing divisions rather than unifying principles. Furthermore, such distinctions can create unnecessary barriers and tensions within the mathematical sciences community by absorbing energy that might be expended more productively. In fact, there are increasing overlaps and beneficial interactions between different areas of the mathematical sciences. . . . [T]he features that unite the mathematical sciences dominate those that divide them.[2]

What is this commonality of experience that is shared across the mathematical sciences? The mathematical sciences aim to understand the world by performing formal symbolic reasoning and computation on abstract structures. One aspect of the mathematical sciences involves unearthing and understanding deep relationships among these abstract structures. Another aspect involves capturing certain features of the world by abstract structures through the process of modeling, performing formal reasoning on these abstract structures or using them as a framework for computation, and then reconnecting back to make predictions about the world—often, this is an iterative process. A related aspect is to use abstract reasoning and structures to make inferences about the world from data. This is linked to the quest to find ways to turn empirical observations into a means to classify, order, and understand reality—the basic promise of science. Through the mathematical sciences, researchers can construct a body of knowledge whose interrelations are understood and where whatever understanding one needs can be found and used. The mathematical sciences also serve as a natural conduit through which concepts, tools, and best practices can migrate from field to field.

A further aspect of the mathematical sciences is to investigate how to make the process of reasoning and computation as efficient as possible and to also characterize their limits. It is crucial to understand that these different aspects of the mathematical sciences do not proceed in isolation from one another. On the contrary, each aspect of the effort enriches the others with new problems, new tools, new insights, and—ultimately—new paradigms.

Put this way, there is no obvious reason that this approach to knowledge should have allowed us to understand the physical world. Yet the entire

---

[2] Engineering and Physical Sciences Research Council (EPSRC), 2010, *International Review of Mathematical Science*. EPSRC, Swindon, U.K., p. 10.

mathematical sciences enterprise has proven not only extraordinarily effective, but indeed essential for understanding our world. This conundrum is often referred to as "the unreasonable effectiveness of mathematics," mentioned in Chapter 2.

In light of that "unreasonable effectiveness," it is even more striking to see, in Figure 3-2, which is analogous to Figure 3-1, how far the mathematical sciences have spread since the Odom report was released in 1998.

Reflecting the reality that underlies Figure 3-2, this report takes a very inclusive definition of "the mathematical sciences." The discipline encompasses the broad range of diverse activities related to the creation and analysis of mathematical and statistical representations of concepts, systems, and processes, whether or not the person carrying out the activity identifies as a

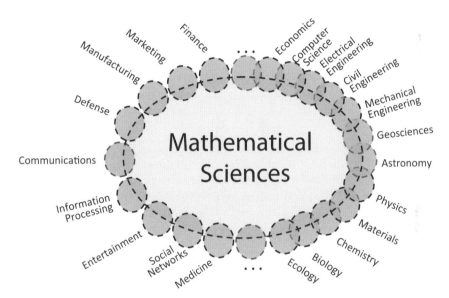

FIGURE 3-2 The mathematical sciences and their interfaces in 2013. The number of interfaces has increased since the time of Figure 3-1, and the mathematical sciences themselves have broadened in response. The academic science and engineering enterprise is suggested by the right half of the figure, while broader areas of human endeavor are indicated on the left. Within the academy, the mathematical sciences are playing a more integrative and foundational role, while within society more broadly their impacts affect all of us—although that is often unappreciated because it is behind the scenes. This schematic is notional, based on the committee's varied and subjective experience rather than on specific data. It does not attempt to represent the many other linkages that exist between academic disciplines and between those disciplines and the broad endeavors on the left, only because the full interplay is too complex for a two-dimensional schematic.

mathematical scientist. The traditional areas of the mathematical sciences are certainly included. But many other areas of science and engineering are deeply concerned with building and evaluating mathematical models, exploring them computationally, and analyzing enormous amounts of observed and computed data. These activities are all inherently mathematical in nature, and there is no clear line to separate research efforts into those that are part of the mathematical sciences and those that are part of computer science or the discipline for which the modeling and analysis are performed.[3] The committee believes the health and vitality of the discipline are maximized if knowledge and people are able to flow easily throughout that large set of endeavors.

So what is the "mathematical sciences community"? It is the collection of people who are advancing the mathematical sciences discipline. Some members of this community may be aligned professionally with two or more disciplines, one of which is the mathematical sciences. (This alignment is reflected, for example, in which conferences they attend, which journals they publish in, which academic degrees they hold, and which academic departments they belong to.) There is great value in the mathematical sciences welcoming these "dual citizens"; their involvement is good for the mathematical sciences, and it enriches the ways in which other fields can approach their work.

The collection of people in the areas of overlap is large. It includes statisticians who work in the geosciences, social sciences, bioinformatics, and other areas that, for historical reasons, became specialized offshoots of statistics. It includes some fraction of researchers in scientific computing and computational science and engineering. It includes number theorists who contribute to cryptography, and real analysts and statisticians who contribute to machine learning. It includes operations researchers, some computer scientists, and physicists, chemists, ecologists, biologists, and economists who rely on sophisticated mathematical science approaches. Some of the engineers who advance mathematical models and computational simulation are also included. It is clear that the mathematical sciences now extend far beyond the definitions implied by the institutions—academic departments, funding sources, professional societies, and principal journals—that support the heart of the field.

As just one illustration of the role that researchers in other fields play in the mathematical sciences, the committee examined public data[4] on National Science Foundation (NSF) grants to get a sense of how much of the research supported by units other than the NSF Division of Mathe-

---

[3] Most of the other disciplines shown in Figure 3-2 also have extensive interactions with other fields, but the full interconnectedness of these endeavors is omitted for clarity.

[4] Available at http://www.nsf.gov/awardsearch/.

matical Sciences (DMS) has resulted in publications that appeared in journals readily recognized as mathematical science ones or that have a title strongly suggesting mathematical or statistical content. While this exercise was necessarily subjective and far from exhaustive, it gave an indication that NSF's support for the mathematical sciences is de facto broader than what is supported by DMS. It also lent credence to the argument that the mathematical sciences research enterprise extends beyond the set of individuals who would traditionally be called mathematical scientists. This exercise revealed the following information:

- Grants awarded over the period 2008-2011 by NSF's Division of Computing and Communication Foundations (part of the Directorate for Computer and Information Science and Engineering) led to 262 publications in the areas of graphs and, to a lesser extent, foundations of algorithms.
- Grants awarded over 2004-2011 by the Division of Physics led to 148 publications in the general area of theoretical physics.
- Grants awarded over 2007-2011 by the Division of Civil, Mechanical, and Manufacturing Innovation in NSF's Engineering Directorate led to 107 publications in operations research.

This cursory examination also counted 15 mathematical science publications resulting from 2009-2010 grants from NSF's Directorate for Biological Sciences. (These publication counts span different ranges of years because the number of publications with apparent mathematical sciences content varies over time, probably due to limited-duration funding initiatives.) For comparison, DMS grants that were active in 2010 led to 1,739 publications. Therefore, while DMS is clearly the dominant NSF supporter of mathematical science research, other divisions contribute in a nontrivial way.

Analogously, membership figures from the Society for Industrial and Applied Mathematics (SIAM) demonstrate that a large number of individuals who are affiliated with academic or industrial departments other than mathematics or statistics nevertheless associate themselves with this mathematical science professional society. Figure 3-3 shows the departmental affiliation of SIAM's nonstudent members.

A recent analysis tried to quantify the size of this community on the interfaces of the mathematical sciences.[5] It found that faculty members in 50 of the top U.S. mathematics departments—who would therefore be in the central disk in Figure 3-2—have published in aggregate some 64,000 research papers since 1971 that have been indexed by Zentralblatt MATH (and thus

---

[5] Joseph Grcar, 2011, Mathematics turned inside out: The intensive faculty versus the extensive faculty. *Higher Education* 61(6): 693-720.

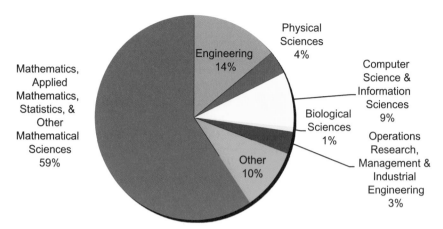

FIGURE 3-3 SIAM members identify the primary department with which they are affiliated. This figure shows the fraction of 6,269 nonstudent members identifying with a particular category.

can be inferred to have mathematical content). Over the same period, some 75,000 research papers indexed by Zentralblatt MATH were published by faculty members in other departments of those same 50 universities. The implication is that a good deal of mathematical sciences research—as much as half of the enterprise—takes place outside departments of mathematics.[6] This also suggests that the scope of most mathematics departments may not mirror the true breadth of the mathematical sciences.

That analysis also created a Venn diagram, reproduced here as Figure 3-4, that is helpful for envisioning how the range of mathematical science research areas map onto an intellectual space that is broader than that covered by most academic mathematics departments. (The diagram also shows how the teaching foci of mathematics and nonmathematics departments differ from their research foci.)

IMPLICATIONS OF THE BROADENING OF
THE MATHEMATICAL SCIENCES

The tremendous growth in the ways in which the mathematical sciences are being used stretches the mathematical science enterprise—its people, teaching, and research breadth. If our overall research enterprise is operat-

---

[6] Some of those 75,000 papers are attributable to researchers in departments of statistics or operations research, which we would clearly count as being in the central disk of Figure 3-2. But the cited paper notes that only about 17 percent of the research indexed by Zentralblatt MATH is classified as dealing with statistics, probability, or operations research.

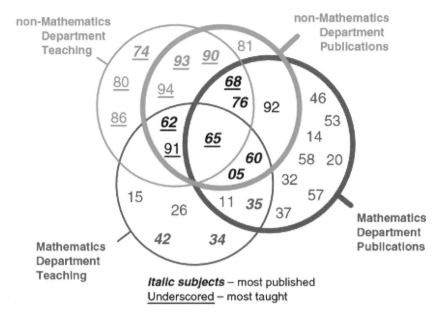

FIGURE 3-4 Representation of the research and teaching span of top mathematics departments and of nonmathematics departments in the same academic institutions. Subjects most published are shown in italics; subjects most taught are underscored. SOURCE: Joseph Grcar, 2011, Mathematics turned inside out: The intensive faculty versus the extensive faculty. *Higher Education* 61(6):693-720, Figure 8. The numbers correspond to the following Zentralblatt MATH classifications:

05 Combinatorics
11 Number theory
14 Algebraic geometry
15 Linear, multilinear algebra
20 Group theory
26 Real functions
32 Several complex variables
34 Ordinary differential equations
35 Partial differential equations
37 Dynamical systems
42 Fourier analysis
46 Functional analysis
53 Differential geometry
57 Manifolds, cell complexes
58 Global analysis

60 Probability theory
62 Statistics
65 Numerical analysis
68 Computer science
74 Mechanics of deformable solids
76 Fluid mechanics
80 Classical thermodynamics
81 Quantum theory
86 Geophysics
90 Operations research
91 Game theory, economics
92 Biology
93 Systems theory, control
94 Information and communications

ing well, the researchers who traditionally call themselves mathematical scientists—the central ellipse in Figure 3-2—are in turn stimulated by the challenges from the frontiers, where new types of phenomena or data stimulate fresh thinking about mathematical and statistical modeling and new technical challenges stimulate deeper questions for the mathematical sciences.

Many people with mathematical sciences training who now work at those frontiers—operations research, computer science, engineering, biology, economics, and so on—have told the committee that they appreciate the grounding provided by their mathematical science backgrounds and that, to them, it is natural and healthy to consider the entire family tree as being a unified whole. Many mathematical scientists and academic math departments have justifiably focused on core areas, and this is natural in the sense that no other community has a mandate to ensure that the core areas remain strong and robust. But it is essential that there be an easy flow of concepts, results, methods, and people across the entirety of the mathematical sciences. For that reason, it is essential that the mathematical sciences community actively embraces the broad community of researchers who contribute intellectually to the mathematical sciences, including people who are professionally associated with another discipline.

Anecdotal information suggests that the number of graduate students receiving training in both mathematics and another field—from biology to engineering—has increased dramatically in recent years. This trend is recognized and encouraged, for example, by the Simons Foundation's Math+X program, which provides cross-disciplinary professorships and support for graduate students and postdoctoral researchers who straddle two fields. If this phenomenon is as general as the committee believes it to be, it shows how mathematic sciences graduate education is contributing to science and engineering generally and also how the interest in interfaces is growing. In order for the community to rationally govern itself, and for funding agencies to properly target their resources, it is necessary to begin gathering data on this trend.

> **Recommendation 3-1: The National Science Foundation should systematically gather data on such interactions—for example, by surveying departments in the mathematical sciences for the number of enrollments in graduate courses by students from other disciplines, as well as the number of enrollments of graduate students in the mathematical sciences in courses outside the mathematical sciences. The most effective way to gather these data might be to ask the American Mathematical Society to extend its annual questionnaires to include such queries.**

Program officers in NSF/DMS and in other funding agencies are aware of many overlaps between the mathematical sciences and other disciplines, and there are many examples of flexibility in funding—mathematical scientists funded by units that primarily focus on other disciplines, and vice versa. DMS in particular works to varying degrees with other NSF units, through formal mechanisms such as shared funding programs and informal mechanisms such as program officers redirecting proposals from one division to another, divisions helping one another in identifying reviewers, and so on. Again, for the mathematical sciences community to have a more complete understanding of its reach, and to help funding agencies best target their programs, the committee recommends that a modest amount of data be collected more methodically.

> **Recommendation 3-2: The National Science Foundation should assemble data about the degree to which research with a mathematical science character is supported elsewhere in the Foundation. (Such an analysis would be of greatest value if it were performed at a level above DMS.) A study aimed at developing this insight with respect to statistical sciences within NSF is under way as this is written, at the request of the NSF assistant director for mathematics and physical sciences. A broader such study would help the mathematical sciences community better understand its current reach, and it could help DMS position its own portfolio to best complement other sources of support for the broader mathematical sciences enterprise. It would provide a baseline for identifying changes in that enterprise over time. Other agencies and foundations that support the mathematical sciences would benefit from a similar self-evaluation.**

Data collected in response to Recommendations 3-1 and 3-2 can help the community, perhaps through its professional societies, adjust graduate training to better reflect actual student behavior. For example, if a significant fraction of mathematics graduate students take courses outside of mathematics, either out of interest or concern about future opportunities, this is something mathematics departments need to know about and respond to. Similarly, a junior faculty member in an interdisciplinary field would benefit by knowing which NSF divisions have funded work in their field. While such knowledge can often be found through targeted conversation, seeing the complete picture would be beneficial for individual researchers, and it might alter the way the mathematical sciences community sees itself.

In a discussion with industry leaders recounted in Chapter 5, the committee was struck by the scale of the demand for workers with mathematical science skills at all degree levels, regardless of their field of training. It heard about the growing demand for people with skills in data analytics, the continuing need for mathematical science skills in the financial sector, and many novel challenges for mathematical scientists working for Internet-driven corporations and throughout the entertainment and gaming sector. There is a burgeoning job market based on mathematical science skills. However, only a small fraction of the people hired by those industry leaders actually hold degrees in mathematics and statistics; these slots are often filled by individuals with training in computer science, engineering, or physical science. While those backgrounds appear to be acceptable to employers, this explosion of jobs based on mathematical science skills represents a great opportunity for the mathematical sciences, and it should stimulate the community in three ways:

- These growing areas of application bring associated research challenges. This is already well recognized in the areas of search technology, financial mathematics, machine learning, and data analytics. No doubt new research challenges will continue to feed back to the mathematical sciences research community as new applications mature;
- The demand for people with appropriate skills will be felt by mathematical science educators, who play a major role in teaching those skills to students in a variety of fields; and
- The large number of career paths now based in the mathematical sciences calls for changes in the curricula for mathematics and statistics undergraduates and graduate students. This will be discussed in the next chapter.

This striking growth in the need for mathematical skills in industry— and for people who might be called "mathematical science practitioners," people with cutting-edge knowledge but who may not be focused on research—presents a great opportunity for the mathematical sciences community. In the past, training in the mathematical sciences was of course essential to the education of researchers in mathematics, statistics, and many fields of science and engineering. And an undergraduate major in mathematics or statistics was always a good basic degree, a stepping-stone to many careers. But the mathematical sciences community tends to view itself as consisting primarily of mathematical science researchers and educators and not extending more broadly. As more people trained in the mathematical sciences at all levels continue in careers that rely on mathematical sciences research, there is an opportunity for the mathematical sciences community to embrace new classes of professionals. At a number of universities, there are opportunities for undergraduate students to engage in research in nonacademic settings and internship programs for graduate students at national laboratories and industry. Some opportunities at both the postdoctoral and more senior levels are available at national laboratories and government agencies. It would be a welcome development for opportunities of this kind to be expanded at all levels. Experiences of this kind at the faculty level can be especially valuable.

In an ideal world, the mathematical sciences community would have a clearer understanding of its scale and impacts. In addition to the steps identified in Recommendations 3-1 and 3-2, annual collection of the following information would allow the community to better understand and improve itself:

- A compilation of important new technology areas, patents, and business start-ups that have applied results from the mathematical sciences, and estimates of employment related to these developments;
- A compilation of existing technology areas that require input and training from the mathematical sciences community, and estimates of employment related to these areas;
- A compilation of new undergraduate and/or graduate programs with significant mathematical science content;
- The ratio of the number of jobs with significant mathematical science content to the number of new graduates (at different levels) in the mathematical sciences;
- An analysis of current employment of people who received federal research or training support in the past, to determine whether they are now in U.S. universities, foreign universities, U.S. or foreign industry, U.S. or foreign government, military, and so on;
- A compilation and analysis of collaborations involving mathematical scientists.

Perhaps the mathematical science professional societies, in concert with some funding agencies, could work to build up such an information base, which would help the enterprise move forward. However, the committee is well aware of the challenges in gathering such data, which would very likely be imprecise and incomplete.

TWO MAJOR DRIVERS OF EXPANSION: COMPUTATION AND BIG DATA

Two factors have combined to spark the enormous expansion of the role of the mathematical sciences: (1) the widespread availability of computing power and the consequent reliance on simulation via mathematical models by so much of today's science, engineering, and technology and (2) the explosion in the amount of data being collected or generated, which often is of a scale that it can only be evaluated through mathematical and statistical techniques. As a result, the two areas of computation and "big data" have emerged as major drivers of mathematical research and the broadening of the enterprise. Needless to say, these are deeply intertwined, and it is becoming increasingly standard for major research efforts to require expertise in both simulation and large-scale data analysis.

Before discussing these two major drivers, it is critical to point out that a great deal of mathematical sciences research continues to be driven by the internal logic of the subject—that is, initiated by individual researchers in response to their best understanding of promising directions. (The often-used phrase "curiosity driven" understates the tremendous effectiveness of this approach over centuries.) In the committee's conversations with many mathematical scientists, it frequently heard the opinion that it is impossible to predict particular research areas in which important new developments are likely to occur. While over the years there have been important shifts in the level of activity in certain subjects—for example, the growing significance of probabilistic methods, the rise of discrete mathematics, and the growing use of Bayesian statistics—the committee did not attempt to exhaustively survey such changes or prognosticate about the subjects that are most likely to produce relevant breakthroughs. The principal lesson is that it continues to be important for funding sources to support excellence wherever it is found and to continue to support the full range of mathematical sciences research.

A new Canadian study on the long-range plan for that country's mathematics and statistics reached a similar conclusion, explaining it as follows:

> [I]t is difficult to predict which particular fields of mathematics and statistics will most guide innovation in the next decades. Indeed, all areas of the mathematical and statistical sciences have the potential to be important to innovation, but the time scale may be very long, and the nature of the link is likely to be surprising. Many areas of the mathematical and statistical sciences that strike us now as abstract and removed from obvious application will be useful in ways that we cannot currently imagine. . . . On the one hand, we need a research landscape that is flexible and non-prescriptive in terms of areas to be supported. We must have a research funding landscape capable of nurturing a broad range of basic and applied research and that can take into account the changing characteristics of the research enterprise itself. And on the other hand, we need to build and maintain infrastructure that will connect the mathematical and statistical sciences to strategic growth areas . . . and will encourage effective technology transfer and innovation across science, industry and society.[7]

---

[7] "Solutions for a Complex Age: Long Range Plan for Mathematical & Statistical Sciences Research in Canada," Natural Sciences & Engineering Research Council (Canada), 2012. Accessed from http://longrangeplan.ca/wp-content/uploads/2012/12/3107_MATH_LRP-1212-web.pdf, January 28, 2013. Quoted text is from p. 24.

Computation

As science, engineering, government, and business rely increasingly on complex computational simulations, it is inevitable that connections between those sectors and the mathematical sciences are strengthened. That is because computational modeling is inherently mathematical. Accordingly, those fields depend on—and profit from—advances in the mathematical sciences and the maintainance of a healthy mathematical science enterprise. The same is true to the extent that those sectors increasingly rely on the analysis of large-scale quantities of data.

This is not to say that a mathematical scientist is needed whenever someone builds or exercises a computer simulation or analyzes data (although the involvement of a mathematical scientist is often beneficial when the work is novel or complex). But it is true that more and more scientists, engineers, and business people require or benefit from higher-level coursework in the mathematical sciences, which strengthens connections between disciplines. And it is also true that the complexity of phenomena that can now be simulated in silico, and the complexity of analyses made possible by terascale data, are pushing research frontiers in the mathematical sciences and challenging those who could have previously learned the necessary skills as they carry out their primary tasks. As this complexity increases, we are finding more and more occasions where specialized mathematical and statistical experience is required or would be beneficial.

Some readers may assume that many of the topics mentioned in this chapter fall in the domain of computer science rather than the mathematical sciences. In fact, many of these areas of inquiry straddle both fields or could be labeled either way. For example, the process of searching data, whether in a database or on the Internet, requires both the products of computer science research and modeling and analysis tools from the mathematical sciences. The challenges of theoretical computer science itself are in fact quite mathematical, and the fields of scientific computing and machine learning sit squarely at the interface of the mathematical sciences and computer science (with insight from the domain of application, in many cases). Indeed, most modeling, simulation, and analysis is built on the output of both disciplines, and researchers with very similar backgrounds can be found in academic departments of mathematics, statistics, or computer science. There is, of course, a great deal of mathematical sciences research that has not that much in common with computer sciences research—and, likewise, a great deal of computer science research that is not particularly close to the mathematical sciences.

Because so many people across science, engineering, and medicine now learn some mathematics and/or statistics, some ask why we need people who specialize in the mathematical sciences to be involved in interdisciplinary teams. The reason is that mathematical science researchers not only create the tools that are translated into applications elsewhere, but they are also the creative partners who can adapt mathematical sciences results appropriately for different problems. This latter sort of collaboration can result in breakthrough capabilities well worth the investment of time that is sometimes associated with establishing a cross-disciplinary team. It is not always enough to rely on the mathematics and statistics that is captured in textbooks or software, for two reasons: (1) progress is continually being made, and off-the-shelf techniques are unlikely to be cutting edge, and (2) solutions tailored to particular situations or questions can often be much more effective than more generic approaches. These are the benefits to the nonmathematical sciences members of the team. For mathematical science collaborators, the benefits are likewise dual: (1) fresh challenges are uncovered that may stimulate new results of intrinsic importance to the mathematical sciences and (2) their mathematical science techniques and insights can have wider impact.

In application areas with well-established mathematical models for phenomena of interest—such as physics and engineering—researchers are able to use the great advances in computing and data collection of recent decades to investigate more complex phenomena and undertake more precise analyses. Conversely, where mathematical models are lacking, the growth in computing power and data now allow for computational simulations using alternative models and for empirically generated relationships as means of investigation.

Computational simulation now guides researchers in deciding which experiments to perform, how to interpret experimental results, which prototypes to build, which medical treatments might work, and so on. Indeed, the ability to simulate a phenomenon is often regarded as a test of our ability to understand it. Over the past 10-15 years, computational capabilities reached a threshold at which statistical methods such as Markov chain Monte Carlo methods and large-scale data mining and analysis became feasible, and these methods have proved to be of great value in a wide range of situations.

For example, at one of its meetings the study committee saw a simulation of biochemical activity developed by Terrence Sejnowski of the Salk Institute for Biological Studies. It was a tour de force of computational simulation—based on cutting-edge mathematical sciences and computer science—that would not have been feasible until recently and that enables novel investigations into complex biological phenomena. As another example, over the past 30 years or so, ultrasound has progressed from providing still images to dynamically showing a beating heart and, more recently, to showing the evolution of a full baby in the womb. The mathematical basis for ultrasound requires solving inverse problems and draws from techniques and results in pure mathematical analysis, in the theory of wave propagation in fluids and elastic media, and in more practical areas such as fast numerical methods and image processing. As ultrasound technologies have improved, new mathematical challenges also need to be addressed.

A recent paper prepared for the Advisory Committee for NSF's Mathematics and Physical Sciences (MPS) Directorate identifies the following five "core elements" that underpin computational science and engineering:[8]

(i)   the development and long-term stewardship of software, including new and "staple" community codes, open source codes, and codes for new or nonconventional architectures;
(ii) the development of models, algorithms, and tools and techniques for verification, validation and uncertainty quantification;
(iii) the development of tools, techniques, and best practices for ultra large data sets;
(iv) the development and adoption of cyber tools for collaboration, sharing, re-use and re-purposing of software and data by MPS communities; and
(v)   education, training and workforce development of the next generation of computational scientists.

The mathematical sciences contribute in essential ways to all the items on this list except the fourth.

The great majority of computational science and engineering can be carried out well by investigators from the field of study: They know how to create a mathematical model of the phenomenon under study, and standard numerical solution tools are adequate. Even in these cases, though, some specialized mathematical insights might be required—such as knowing how and when to add "artificial viscosity" to a representation of fluid flow, or how to handle "stiffness" in sets of ordinary differential equations—but that degree of skill has spread among numerical modelers in many disciplines. However, as the phenomena being modeled become increasingly complex, perhaps requiring specialized articulation between models at different scales and of different mathematical types, specialized mathematical science skills become more and more important. Absent such skills and experience, computational models can be unstable or even produce unreliable results. Validation of such complex models requires very specialized experience, and the critical task of quantifying their uncertainties can be

---

[8] Sharon Glotzer, David Keyes, Joel Tohline, and Jerzy Leszczynski, "Computational Science." May 14, 2010. Available at http://www.nsf.gov/attachments/118651/public/MPSAC_ Computational_Science_White_Paper.pdf.

very challenging.[9] And when models are developed empirically by finding patterns in large-scale data—which is becoming ever more common—there are many opportunities for spurious associations. The research teams must have strong statistical skills in order to create reliable knowledge in such cases.

In response to the need to harness this vast computational power, the community of mathematical scientists who are experts in scientific computation continues to expand. This cadre of researchers develops improved solution methods, algorithms for gridding schemes and computational graphics, and so on. Much more of that work will be stimulated by new computer architectures now emerging. And, a much broader range of mathematical science challenges stem from this trend. The theory of differential equations, for example, is challenged to provide structures that enable us to analyze approximations to multiscale models; stronger methods of model validation are needed; algorithms need to be developed and characterized; theoretical questions in computer science need resolution; and so on. Though often simply called "software," the steps to represent reality on a computer pose a large number of challenges for the mathematical sciences.[10]

A prime example of how expanding computational and data resources have led to the "mathematization" of a field of science is the way that biology became much more quantitative and dependent on mathematical and statistical modeling following the emergence of genomics. High-throughput data in biology has been an important driver for new statistical research over the past 10-15 years. Research in genomics and proteomics relies heavily on the mathematical sciences, often in challenging ways, and studies of disease, evolution, agriculture, and other topics have consequently become quantitative as genomic and proteomic information is incorporated as a foundation for research. Arguably, this development has placed statisticians as central players in one of the hottest fields of science. Over the next 10-15 years, acquiring genomic data will become fairly straightforward, and increasingly it will be available to illuminate biological processes. A

---

[9] See Kristin Sainani, 2011, Error!–What biomedical computing can learn from its mistakes. *Biomedical Computation Review*, September 1, gives an overview of sources of error and points to some striking published studies. Available at http://biomedicalcomputationreview. org/content/error-%E2%80%93-what-biomedical-computing-can-learn-its-mistakes. See also National Research Council, 2012. *Assessing the Reliability of Complex Models: Mathematical and Statistical Foundations of Verification, Validation, and Uncertainty Quantification*. The National Academies Press, Washington, D.C.

[10] The full range of mathematical sciences challenges associated with advanced computing for science and engineering is discussed in Chapter 6 of NRC, 2008, *The Potential Impact of High-End Capacity Computing on Four Illustrative Fields of Science and Engineering*, The National Academies Press, Washington, D.C.

lot of that work depends on insights from statistics, but mathematics also provides foundations: For example, graph theoretic methods are essential in evolutionary biology; new algorithms from discrete mathematics and computer science play essential roles in search, comparison, and knowledge extraction; dynamical systems theory plays an important role in ecology; and more traditional applied mathematics is commonly used in computational neuroscience and systems biology. As biology transitions from a descriptive science to a quantitative science, the mathematical sciences will play an enormous role.

To different degrees, the social sciences are also embracing the tools of the mathematical sciences, especially statistics, data analytics, and mathematics embedded in simulations. For example, statistical models of disease transmission have provided very valuable insights about patterns and pathways. Business, especially finance and marketing, is increasingly dependent on methods from the mathematical sciences. Some topics in the humanities have also benefited from mathematical science methods, primarily data mining, data analysis, and the emerging science of networks.

The mathematical sciences are increasingly contributing to data-driven decision making in health care. Operations research is being applied to model the processes of health care delivery so they can be methodically improved. The applications use different forms of simulation, discrete optimization, Markov decision processes, dynamic programming, network modeling, and stochastic control. As health care practices move to electronic health care records, enormous amounts of data are becoming available and in need of analysis; new methods are needed because these data are not the result of controlled trials. The new field of comparative effectiveness research, which relies a great deal on statistics, aims to build on data of that sort to characterize the effectiveness of various medical interventions and their value to particular classes of patients.

"Big Data"

Embedded in several places in this discussion is the observation that data volumes are exploding, placing commensurate demands on the mathematical sciences. This prospect has been mentioned in a large number of previous reports about the discipline, but it has become very real in the past 15 years or so. A sign that this area has come of age is the unveiling in March 2012 by the White House Office of Science and Technology Policy (OSTP) of the Big Data Research and Development Initiative. OSTP Director John Holdren said pointedly that "it's not the data per se that create value. What really matters is our ability to derive from them new insights, to recognize relationships, to make increasingly accurate predictions. Our ability, that is, to move from data, to knowledge, to action." It is precisely in the realm of moving from raw data via analysis to knowledge that the mathematical sciences are essential. Large, complex data sets and data streams play a significant role in stimulating new research applications across the mathematical sciences, and mathematical science advances are necessary to exploit the value in these data.

However, the role of the mathematical sciences in this area is not always recognized. Indeed, the stated goals for the OSTP initiative,

> to advance state-of-the-art core technologies needed to collect, store, preserve, manage, analyze, and share huge quantities of data; harness these technologies to accelerate the pace of discovery in science and engineering; strengthen our national security; and transform teaching and learning; and to expand the workforce needed to develop and use Big Data technologies,

seem to understate the amount of intellectual effort necessary to actually enable the move from data, to knowledge, to action.

Multiple issues of fundamental methodology arise in the context of large data sets. Some arise from the basic issue of scalability—that techniques developed for small or moderate-sized data sets do not translate to modern massive data sets—or from problems of data streaming, where the data set is changing while the analysis goes on. Typical questions include how to increase the signal-to-noise ratio in the recorded data, how to detect a new or different state quickly (anomaly detection), algorithmic/data structure solutions that allow fast computation of familiar statistics, hardware solutions that allow efficient or parallel computations. Data that are high-dimensional pose new challenges: New paradigms of statistical inference arise from the exploratory nature of understanding large complex data sets, and issues arise of how best to model the processes by which large, complex data sets are formed. Not all data are numerical—some are categorical, some are qualitative, and so on—and mathematical scientists contribute perspectives and techniques for dealing with both numerical and non-numerical data, and with their uncertainties. Noise in the data-gathering process needs to be modeled and then—where possible—minimized; a new algorithm can be as powerful an enhancement to resolution as a new instrument. Often, the data that can be measured are not the data that one ultimately wants. This results in what is known as an inverse problem—the process of collecting data has imposed a very complicated transformation on the data one wants, and a computational algorithm is needed to invert the process. The classic example is radar, where the shape of an object is reconstructed from how radio waves bounce off it. Simplifying the data so as to find its underlying structure is usually essential in large data sets. The general goal of dimensionality reduction—taking data with a large number of measurements and finding which combinations of the measurements are sufficient to embody the essential features of the data set—is pervasive. Various methods with their roots in linear algebra and statistics are used and continually being improved, and increasingly deep results from real analysis and probabilistic methods—such as random projections and diffusion geometry—are being brought to bear.

Statisticians contribute a long history of experience in dealing with the intricacies of real-world data—how to detect when something is going wrong with the data-gathering process, how to distinguish between outliers that are important and outliers that come from measurement error, how to design the data-gathering process so as to maximize the value of the data collected, how to cleanse the data of inevitable errors and gaps. As data sets grow into the terabyte and petabyte range, existing statistical tools may no longer suffice, and continuing innovation is necessary. In the realm of massive data, long-standing paradigms can break—for example, false positives can become the norm rather than the exception—and more research endeavors need strong statistical expertise.

For example, in a large portion of data-intensive problems, observations are abundant and the challenge is not so much how to avoid being deceived by a small sample size as to be able to detect relevant patterns. As noted in the *New York Times*: "In field after field, computing and the Web are creating new realms of data to explore sensor signals, surveillance tapes, social network chatter, public records and more."[11] This may call for researchers in machine or statistical learning to develop algorithms that predict an outcome based on empirical data, such as sensor data or streams from the Internet. In that approach, one uses a sample of the data to discover relationships between a quantity of interest and explanatory variables. Strong mathematical scientists who work in this area combine best practices in data modeling, uncertainty management, and statistics, with insight about the application area and the computing implementation. These prediction problems arise everywhere: in finance and medicine, of course, but they are also crucial to the modern economy so much so that businesses like Netflix, Google, and Facebook rely on progress in this area. A recent trend is that statistics graduate students who in the past often ended up in pharmaceutical companies, where they would design clinical trials, are increasingly also being recruited by companies focused on Internet commerce.

Finding what one is looking for in a vast sea of data depends on search algorithms. This is an expanding subject, because these algorithms need to search a database where the data may include words, numbers, images and video, sounds, answers to questionnaires, and other types of data, all linked

---

[11] Steve Lohr, 2009, For today's graduate, just one word: Statistics. *New York Times*, August 5.

together in various ways. It is also necessary for these methods to become increasingly "intelligent" as the scale of the data increases, because it is insufficient to simply identify matches and propose an ordered list of hits. New techniques of machine learning continue to be developed to address this need. Another new consideration is that data often come in the form of a network; performing mathematical and statistical analyses on networks requires new methods.

Statistical decision theory is the branch of statistics specifically devoted to using data to enable optimal decisions. What it adds to classical statistics beyond inference of probabilities is that it integrates into the decision information about costs and the value of various outcomes. It is critical to many of the projects envisioned in OSTP's Big Data initiative. To pick two examples from many,[12] the Center for Medicare and Medicaid Services (CMS) has the program Using Administrative Claims Data (Medicare) to Improve Decision-Making while the Food and Drug Administration is establishing the Virtual Laboratory Environment, which will apply "advanced analytical and statistical tools and capabilities, crowd-sourcing of analytics to predict and promote public health." Better decision making has been singled out as one of the most promising ways to curtail rising medical costs while optimizing patient outcomes, and statistics is at the heart of this issue.

Ideas from statistics, theoretical computer science, and mathematics have provided a growing arsenal of methods for machine learning and statistical learning theory: principal component analysis, nearest neighbor techniques, support vector machines, Bayesian and sensor networks, regularized learning, reinforcement learning, sparse estimation, neural networks, kernel methods, tree-based methods, the bootstrap, boosting, association rules, hidden Markov models, and independent component analysis—and the list keeps growing. This is a field where new ideas are introduced in rapid-fire succession, where the effectiveness of new methods often is markedly greater than existing ones, and where new classes of problems appear frequently.

Large data sets require a high level of computational sophistication because operations that are easy at a small scale—such as moving data between machines or in and out of storage, visualizing the data, or displaying results—can all require substantial algorithmic ingenuity. As a data set becomes increasingly massive, it may be infeasible to gather it in one place and analyze it as a whole. Thus, there may be a need for algorithms that operate in a distributed fashion, analyzing subsets of the data and aggregating those results to understand the complete set. One aspect of this is the

---

[12] The information and quotation here are drawn from OSTP, "Fact Sheet: Big Data Across the Federal Government, March 29, 2012." Available at http://www.whitehouse.gov/sites/default/files/microsites/ostp/big_data_fact_sheet_final.pdf.

challenge of data assimilation, in which we wish to use new data to update model parameters without reanalyzing the entire data set. This is essential when new waves of data continue to arrive, or subsets are analyzed in isolation of one another, and one aims to improve the model and inferences in an adaptive fashion—for example, with streaming algorithms. The mathematical sciences contribute in important ways to the development of new algorithms and methods of analysis, as do other areas as well.

Simplifying the data so as to find their underlying structure is usually essential in large data sets. The general goal of dimensionality reduction— taking data with a large number of measurements and finding which combinations of the measurements are sufficient to embody the essential features of the data set—is pervasive. Various methods with their roots in linear algebra, statistics, and, increasingly, deep results from real analysis and probabilistic methods—such as random projections and diffusion geometry—are used in different circumstances, and improvements are still needed. Such issues are central to NSF's Core Techniques and Technologies for Advancing Big Data Science and Engineering program and to data as diverse as those from climate, genomics, and threat reduction. Related to search and also to dimensionality reduction is the issue of anomaly detection—detecting which changes in a large system are abnormal or dangerous, often characterized as the needle-in-a-haystack problem. The Defense Advanced Research Projects Agency (DARPA) has its Anomaly Detection at Multiple Scales program on anomaly-detection and characterization in massive data sets, with a particular focus on insider-threat detection, in which anomalous actions by an individual are detected against a background of routine network activity. A wide range of statistical and machine learning techniques can be brought to bear on this, some growing out of statistical techniques originally used for quality control, others pioneered by mathematicians in detecting credit card fraud.

Two types of data that are extraordinarily important yet exceptionally subtle to analyze are words and images. The fields of text mining and natural language processing deal with finding and extracting information and knowledge from a variety of textual sources, and creating probabilistic models of how language and grammatical structures are generated. Image processing, machine vision, and image analysis attempt to restore noisy image data into a form that can be processed by the human eye, or to bypass the human eye altogether and understand and represent within a computer what is going on in an image without human intervention.

Related to image analysis is the problem of finding an appropriate language for describing shape. Multiple approaches, from level sets to "eigenshapes," come in, with differential geometry playing a central role. As part of this problem, methods are needed to describe small deformations of shapes, usually using some aspect of the geometry of the space of diffeomorphisms. Shape analysis also comes into play in virtual surgery, where surgical outcomes are simulated on the computer before being tried on a patient, and in remote surgery for the battlefield. Here one needs to combine mathematical modeling techniques based on the differential equations describing tissue mechanics with shape description and visualization methods.

As our society is learning somewhat painfully, data must be protected. The need for privacy and security has given rise to the areas of privacy-preserving data mining and encrypted computation, where one wishes to be able to analyze a data set without compromising privacy, and to be able to do computations on an encrypted data set while it remains encrypted.

CONTRIBUTIONS OF MATHEMATICAL SCIENCES TO SCIENCE AND ENGINEERING

The mathematical sciences have a long history of interaction with other fields of science and engineering. This interaction provides tools and insights to help those other fields advance; at the same time, the efforts of those fields to push research frontiers routinely raise new challenges for the mathematical sciences themselves. One way of evaluating the state of the mathematical sciences is to examine the richness of this interplay. Some of the interactions between mathematics and physics are described in Chapter 2, but the range extends well beyond physics. A compelling illustration of how much other fields rely on the mathematical sciences arises from examining those fields' own assessments of promising directions and identifying the directions that are dependent on parallel progress in the mathematical sciences. A number of such illustrations have been collected in Appendix D.

CONTRIBUTIONS OF MATHEMATICAL SCIENCES TO INDUSTRY

The role of the mathematical sciences in industry has a long history, going back to the days when the Egyptians used the 3-4-5 right triangle to restore boundaries of farms after the annual flooding of the Nile. That said, the recent period is one of remarkable growth and diversification. Even in old-line industries, the role of the mathematical sciences has expanded. For example, whereas the aviation industry has long used mathematics in the design of airplane wings and statistics in ensuring quality control in production, now the mathematical sciences are also crucial to GPS and navigation systems, to simulating the structural soundness of a design, and to optimizing the flow of production. Instead of being used just to streamline cars and model traffic flows, the mathematical sciences are also involved in the latest developments, such as design of automated vehicle detection and avoidance systems that may one day lead to automated driving. Whereas statistics has long been a key element of medical trials, now the mathematical sciences are involved in drug design and in modeling new ways for drugs to be delivered to tumors, and they will be essential in making inferences in circumstances that do not allow double-blind, randomized clinical trials. The financial sector, which once relied on statistics to design portfolios that minimized risk for a given level of return, now makes use of statistics, machine learning, stochastic modeling, optimization, and the new science of networks in pricing and designing securities and in assessing risk.

What is most striking, however, is the number of new industries that the mathematical sciences are a part of, often as a key enabler. The encryption industry makes use of number theory to make Internet commerce possible. The "search" industry relies on ideas from the mathematical sciences to make the Internet's vast resources of information searchable. The social networking industry makes use of graph theory and machine learning. The animation and computer game industry makes use of techniques as diverse as differential geometry and partial differential equations. The biotech industry heavily uses the mathematical sciences in modeling the action of drugs, searching genomes for genes relevant to human disease or relevant to bioengineered organisms, and discovering new drugs and understanding how they might act. The imaging industry uses ideas from differential geometry and signal processing to procure minimally invasive medical and industrial images and, within medicine, adds methods from inverse problems to design targeted radiation therapies and is moving to incorporate the new field of computational anatomy to enable remote surgery. The online advertising industry uses ideas from game theory and discrete mathematics to price and bid on online ads and methods from statistics and machine learning to decide how to target those ads. The marketing industry now employs sophisticated statistical and machine learning techniques to target customers and to choose locations for new stores. The credit card industry uses a variety of methods to detect fraud and denial-of-service attacks. Political campaigns now make use of complex models of the electorate, and election-night predictions rely on integrating these models with exit polls. The semiconductor industry uses optimization to design computer chips and in simulating the manufacture and behavior of designer materials. The mathematical sciences are now present in almost every industry, and the range of mathematical sciences being used would have been unimaginable a generation ago.

This point is driven home by the following list of case studies assembled for the SIAM report *Mathematics in Industry*.[13] This list is just illustrative, but its breadth is striking:

---

[13] Society for Industrial and Applied Mathematics, 2012, *Mathematics in Industry*. SIAM, Philadelphia, Pa.

- Predictive analytics,
- Image analysis and data mining,
- Scheduling and routing of deliveries,
- Mathematical finance,
- Algorithmic trading,
- Systems biology,
- Molecular dynamics,
- Whole-patient models,
- Oil basin modeling,
- Virtual prototyping,
- Molecular dynamics for product engineering,
- Multidisciplinary design optimization and computer-aided design,
- Robotics,
- Supply chain management,
- Logistics,
- Cloud computing,
- Modeling complex systems,
- Viscous fluid flow for computer and television screen design,
- Infrastructure management for smart cities, and
- Computer systems, software, and information technology.

The reader is directed to the SIAM report to see the details of these case studies,[14] which provide many examples of the significant and cost-effective impact of mathematical science expertise and research on innovation, economic competitiveness, and national security.

Another recent report on the mathematical sciences in industry came to the following conclusions:

> It is evident that, in view of the ever-increasing complexity of real life applications, the ability to effectively use mathematical modelling, simulation, control and optimisation will be the foundation for the technological and economic development of Europe and the world.[15]

> Only [the mathematical sciences] can help industry to optimise more and more complex systems with more and more constraints.[16]

However, that report also points out the following truism:

> [Engineering] designers use virtual design environments that rely heavily on mathematics, and produce new products that are well recognised by

---

[14] Ibid., pp. 9-24.

[15] European Science Foundation, 2010, *Mathematics and Industry*. Strasbourg, France, p. 8.

[16] Ibid., p. 12.

management. The major effort concerned with the construction of reliable, robust and efficient virtual design environments is, however, not recognised. As a result, mathematics is not usually considered a key technology in industry. The workaround for this problem usually consists of leaving the mathematics to specialised small companies that often build on mathematical and software solutions developed in academia. Unfortunately, the level of communication between these commercial vendors and their academic partners with industry is often at a very low level. This, in turn, leads to the observation that yesterday's problems in industry can be solved, but not the problems of today and tomorrow. The latter can only be addressed adequately if an effort is made to drastically improve the communication between industrial designers and mathematicians.[17]

One way to address this communication challenge, of course, is to include high-caliber mathematical scientists within corporate R&D units.

The mathematical functions of greatest value in these and other successful applications were characterized by R&D managers in 1996 as follows:[18]

- Modeling and simulation,
- Mathematical formulation of problems,
- Algorithm and software development,
- Problem-solving,
- Statistical analysis,
- Verifying correctness, and
- Analysis of accuracy and reliability.

To this list, the European Science Foundation report adds optimization, noting (p. 12) that "due to the increased computational power and the achievements obtained in speed-up of algorithms . . . optimization of products has [come] into reach. This is of vital importance to industry." This is a very important development, and it opens up new challenges for the mathematical sciences, such as how to efficiently explore design options and how to characterize the uncertainties of this computational sampling of the space.

These sorts of opportunities were implicitly recognized in a report from the Chinese Academy of Sciences (CAS), *Science and Technology in China: A Roadmap to 2050.*[19] That report identified eight systems of importance for socioeconomic development: sustainable energy and resources,

---

[17] Ibid., p. 9.

[18] Society for Industrial and Applied Mathematics, 1996, *Mathematics in Industry.* Available at http://www.siam.org/reports/mii/1996/listtables.php#lt4.

[19] CAS, 2010, *Science & Technology in China: A Roadmap to 2050.* Springer. Available online at http://www.bps.cas.cn/ztzl/cx2050/nrfb/201008/t20100804_2917262.html.

advanced materials and intelligent manufacturing, ubiquitous information networking, ecological and high-value agriculture and biological industry, health assurance, ecological and environmental conservation, space and ocean exploration, and national and public security. To support the maturation of these systems, the report goes on to identify 22 science and technology initiatives. Of these, three will build on the mathematical sciences: an initiative to create "ubiquitous informationized manufacturing system[s]," another to develop exascale computing technology, and a third in basic cross-disciplinary research in mathematics and complex systems.

The last-mentioned initiative is intended to research the "basic principles behind various kinds of complex systems," and the report recommends that major efforts be made in the following research directions:

- Mathematical physics equations;
- Multiscale modeling and computation of complex systems;
- Machine intelligence and mathematics mechanization;
- Theories and methods for stochastic structures and data;
- Collective behaviors of multiagent complex systems, their control and intervention; and
- Complex stochastic networks, complex adaptive systems, and related areas.

In particular, the report recommends that due to the fundamental importance of complex systems, the Chinese government should provide sustained and steady support for research into such systems so as to achieve major accomplishments in this important field.

One more example of the role of the mathematical sciences in industry comes from the NRC report *Visionary Manufacturing Challenges for 2020,*[20] which identified R&D that would be necessary to advance national capabilities in manufacturing. A number of these capabilities rely on research in modeling and simulation, control theory, and informatics:

- Ultimately, simulations of manufacturing systems would be based on a unified taxonomy for process characteristics that include human characteristics in process models. Other areas for research include a general theory for adaptive systems that could be translated into manufacturing processes, systems, and the manufacturing enterprise; tools to optimize design choices to incorporate the most affordable manufacturing approaches; and systems research on the interaction between workers and manufacturing processes for the development of adaptive, flexible controls. (p. 39)

---

[20] NRC, 1998, *Visionary Manufacturing Challenges for 2020.* National Academy Press, Washington, D.C.

- Modeling and simulation capabilities for evaluating process and enterprise scenarios will be important in the development of reconfigurable enterprises. . . . Virtual prototyping of manufacturing processes and systems will enable manufacturers to evaluate a range of choices for optimizing their enterprises. Promising areas for the application of modeling and simulation technology for reconfigurable systems include neural networks for optimizing reconfiguration approaches and artificial intelligence for decision making. . . . Processes that can be adapted or readily reconfigured will require flexible sensors and control algorithms that provide precision process control of a range of processes and environments. (pp. 39-40)
- Research in enterprise modeling tools will include 'soft' modeling (e.g., models that consider human behavior as an element of the system and models of information flow and communications), the optimization and integration of mixed models, the optimization of hardware systems, models of organizational structures and cross-organizational behavior, and models of complex or nonlinear systems and processes. (p. 44)
- Future information systems will have to be able to collect and sift through vast amounts of information. (p. 44)

Today's renewed emphasis on advanced manufacturing, as exemplified by the recent report *Capturing Domestic Competitive Advantage in Advanced Manufacturing*,[21] also relies implicitly on advances in the mathematical sciences. The 11 cross-cutting technology areas identified in that report (p. 18) as top candidates for R&D investments rely in multiple ways on modeling, simulation, and analysis of complex systems, analysis of large amounts of data, control, and optimization.

CONTRIBUTIONS OF MATHEMATICAL SCIENCES TO NATIONAL SECURITY

National security is another area that relies heavily on the mathematical sciences. The National Security Agency (NSA), for example, employs roughly 1,000 mathematical scientists, although the number might be half that or twice that depending on how one defines such scientists.[22] They include people with backgrounds in core and applied mathematics, probability, and statistics, but people with computer science backgrounds are not included in that count. NSA hires some 40-50 mathematicians per year, and it tries to keep that rate steady so that the mathematical sciences com-

---

[21] White House, Office of Science and Technology Policy, 2012.

[22] Alfred Hales, former head of the Institute for Defense Analyses, Center for Communications Research–La Jolla, presentation to the committee in December 2010. IDA's La Jolla center conducts research for the NSA.

munity knows it can depend on that level of hiring. The NSA is interested in maintaining a healthy mathematical sciences community in the United States, including a sufficient supply of well-trained U.S. citizens. Most of NSA's mathematical sciences hiring is at the graduate level, about evenly split between the M.S. and Ph.D. levels. The agency hires across almost all fields of the mathematical sciences rather than targeting specific subfields, because no one can predict the mix of skills that will be important over an employee's decades-long career. For example, few mathematicians would have guessed decades ago that elliptic curves would be of vital interest to NSA, and now they are an important specialty underlying cryptology.

While cryptology is explicitly dependent on mathematics, many other links exist between the mathematical sciences and national security. One example is analysis of networks (discussed in the next section), which is very important for national defense. Another is scientific computing. One of the original reasons for John von Neumann's interest in creating one of the first computers was to be able to do the computations necessary to simulate what would happen inside a hydrogen bomb. Years later, with atmospheric and underground nuclear testing banned, the country relies once again on simulations, this time to maintain the readiness of its nuclear arsenal. Because national defense relies in part on design and manufacturing of cutting-edge equipment, it also relies on the mathematical sciences through their contributions to advanced engineering and manufacturing. The level of sophistication of these tools has ratcheted steadily upward. The mathematical sciences are also essential to logistics, simulations used for training and testing, war-gaming, image and signal analysis, control of satellites and aircraft, and test and evaluation of new equipment. Figure 3-5, reproduced from *Fueling Innovation and Discovery: The Mathematical Sciences in the 21st Century,*[23] captures the broad range of ways in which the mathematical sciences contribute to national defense.

New devices, on and off the battlefield, have come on stream and furnish dizzying quantities of data, more than can currently be analyzed. Devising ways to automate the analysis of these data is a highly mathematical and statistical challenge. Can a computer be taught to make sense of a satellite image, detecting buildings and roads and noticing when there has been a major change in the image of a site that is not due to seasonal variation? How can one make use of hyperspectral data, which measure light reflected in all frequencies of the spectrum, in order to detect the smoke plume from a chemical weapons factory? Can one identify enemy vehicles and ships in a cluttered environment? These questions and many others are inherently dependent on advances in the mathematical sciences.

---

[23] National Research Council, 2012. The National Academies Press, Washington, D.C.

A very serious threat that did not exist in earlier days is that crucial networks are constantly subject to sophisticated attacks by thieves, mischief-makers, and hackers of unknown origin. Adaptive techniques based on the mathematical sciences are essential for reliable detection and prevention of such attacks, which grow in sophistication to elude every new strategy for preventing them.

The Department of Defense has adopted seven current priority areas for science and technology investment to benefit national security.

1. *Data to decisions.* Science and applications to reduce the cycle time and manpower requirements for analysis and use of large data sets.
2. *Engineered resilient systems.* Engineering concepts, science, and design tools to protect against malicious compromise of weapon systems and to develop agile manufacturing for trusted and assured defense systems.
3. *Cyber science and technology.* Science and technology for efficient, effective cyber capabilities across the spectrum of joint operations.
4. *Electronic warfare/electronic protection.* New concepts and technology to protect systems and extend capabilities across the electromagnetic spectrum.
5. *Countering weapons of mass destruction (WMD):* Advances in DOD's ability to locate, secure, monitor, tag, track, interdict, eliminate, and attribute WMD weapons and materials.
6. *Autonomy.* Science and technology to achieve autonomous systems that reliably and safely accomplish complex tasks in all environments.
7. *Human systems.* Science and technology to enhance human-machine interfaces, increasing productivity and effectiveness across a broad range of missions.[24]

While the mathematical sciences are clearly of importance to the first and third of these priority areas, they also have key roles to play in support of all of the others. Advances in the mathematical sciences that allow simulation-based design, testing, and control of complex systems are essential for creating resilient systems. Improved methods of signal analysis and processing, such as faster algorithms and more sensitive schemes for pattern recognition, are needed to advance electronic warfare and protection. Rapidly developing tools for analyzing social networks, which are based on novel methods of statistical analysis of networks, are being applied in order

---

[24] Department of Defense, 2012, Memorandum on *Science and Technology (S&T) Priorities for Fiscal Years 2013-17 Planning.* Available at http://acq.osd.mil/chieftechnologist/publications/docs/OSD%2002073-11.pdf. Accessed May 3, 2012.

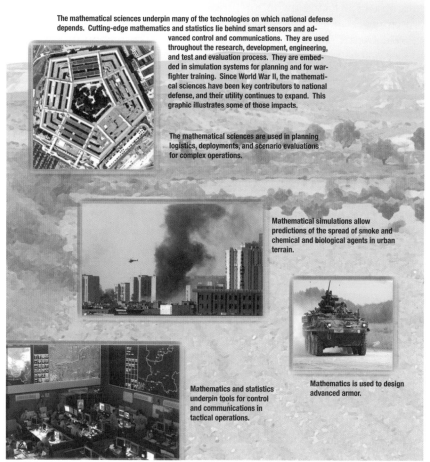

Mathematical Sciences Inside...

The mathematical sciences underpin many of the technologies on which national defense depends. Cutting-edge mathematics and statistics lie behind smart sensors and advanced control and communications. They are used throughout the research, development, engineering, and test and evaluation process. They are embedded in simulation systems for planning and for warfighter training. Since World War II, the mathematical sciences have been key contributors to national defense, and their utility continues to expand. This graphic illustrates some of those impacts.

The mathematical sciences are used in planning logistics, deployments, and scenario evaluations for complex operations.

Mathematical simulations allow predictions of the spread of smoke and chemical and biological agents in urban terrain.

Mathematics and statistics underpin tools for control and communications in tactical operations.

Mathematics is used to design advanced armor.

FIGURE 3-5 Mathematical sciences inside the battlefield.

The Battlefield

Signal analysis and control theory are essential for drones.

Large-scale computational codes are used to design aircraft, simulate flight paths, and train personnel.

Signal processing facilitates communication capabilities.

Mobile translation systems employ voice recognition software to reduce language barriers when human linguists are not available. More generally, math-based simulations are used in mission and specialty training.

Satellite-guided weapons utilize GPS for highly-precise targeting, while mathematical methods improve ballistics.

Modeling and simulation facilitates trade-off analysis during vehicle design, while statistics underpins test and evaluation.

to improve our capabilities in counter-WMD. And mathematical science methods of machine learning are necessary for improving our capabilities in autonomy and human-machine interfaces. Computational neuroscience, which relies heavily on the mathematical sciences, is also a promising area for future developments in human-machine interfaces.

The realm of threat detection in general requires a multiplicity of techniques from the mathematical sciences. How can one most quickly detect patterns that might indicate the spread of a pathogen spread by bioterrorism? How does one understand the structure of a terrorist network? Can one design a power grid or a transportation network in such a way that it is maximally resistant to attack? A new threat, cyberwarfare, grows in magnitude. Fighting back will involve a multipronged response: better encryption, optimized design of networks, and the burgeoning mathematics- and statistics-intensive field of anomaly detection.

4

Important Trends in the Mathematical Sciences

This chapter draws on the inputs gathered by the committee at its meetings and on the members' collective experiences to identify trends that are affecting the mathematical sciences. These trends are likely to continue, and they call for adjustments in the way the enterprise—including individual professionals, academic departments, university administrators, professional societies, and funding agencies—supports the discipline and how the community adjusts for the needs between now and 2025. Recommendations on the necessary adjustments are included as appropriate.

INCREASING IMPORTANCE OF CONNECTIONS FOR MATHEMATICAL SCIENCES RESEARCH

Based on testimony received at its meetings, conference calls with leading researchers (see Appendix B), and the experiences of its members, the committee concludes that the importance of connections among areas of research has been growing over the past two decades or more. This trend has been accelerating over the past 10-15 years, and all indications are that connections will continue to be very important in the coming years. Connections are of two types:

- The discipline itself—research that is internally motivated—is growing more strongly interconnected, with an increasing need for research to tap into two or more fields of the mathematical sciences;
- Research that is motivated by, or applied to, another field of science, engineering, business, or medicine is expanding, in terms of

both the number of fields that now have overlaps with the mathematical sciences and the number of opportunities in each area of overlap.

The second of these trends was discussed in Chapter 3. This section elaborates on the trend toward greater connectivity on the part of the mathematical sciences themselves.

Internally driven research within the mathematical sciences is showing an increasing amount of collaboration and research that involves two or more fields within the discipline. Some of the most exciting advances have built on fields of study—for example, probability and combinatorics—that had not often been brought together in the past. This change is nontrivial because large bodies of knowledge must be internalized by the investigator(s).

The increased interconnectivity of the mathematical sciences community has led, as one would expect, to an increase in joint work. To cite just one statistic, the average number of authors per paper in the *Annals of Mathematics* has risen steadily, from 1.2 in the 1960s to 1.8 in the 2000s. While this increase is modest compared to the multiauthor traditions in many fields, it is significant because it shows that the core mathematics covered by this leading journal is trending away from the solitary researcher model that is embedded in the folklore of mathematics. It also suggests what has long been the experience of leading mathematicians: that the various subfields in mathematics depend on one another in ways that are unpredictable but almost inevitable, and so more individuals need to collaborate in order to bring all the necessary expertise to bear on today's problems.

While some collaborative work involves mathematical scientists of similar backgrounds joining forces to attack a problem of common interest, in other cases the collaborators bring complementary backgrounds. In such cases, the increased collaboration in recent years has led to greater cross-fertilization of fields—to ideas from one field being used in another to make significant advances. A few recent examples of this interplay among fields are given here. This list is certainly not exhaustive, but it indicates the vitality of cross-disciplinary work and its importance in modern mathematical sciences.

Example 1: Cross-Fertilization in Core Mathematics

Recent years have seen major striking examples of ideas and results from one field of core mathematics imported to establish important results in other fields. An example is the resolution of the Poincaré conjecture, the most famous problem in topology, by using ideas from geometry and analysis along with results about a class of abstract metric spaces. As another example, there is mounting evidence that the same geometric flow techniques used in resolving the Poincaré conjecture can be applied to complex algebraic manifolds to understand the existence and nonexistence of canonical metrics—for example, the Kähler-Einstein metric—on these manifolds. Recently, ideas from algebraic topology dating back to the 1960s, A_∞-algebras and modules, have been used in the study of invariants of symplectic manifolds and low-dimensional topological manifolds. These lead to the introduction of ever more sophisticated and powerful algebraic structures in the study of these topological and geometric problems.

In a different direction, deep connections have recently been discovered between random matrix theory, combinatorics, and number theory. The zeros of the Riemann zeta function seem to follow—with astounding precision—a distribution connected with the eigenvalues of large random matrices. This distribution was originally studied as a way of understanding the spectral lines of heavy atoms. The same distribution occurs in many other areas—for example with respect to the standard tableaux of combinatorics—and in the study of quantum chaos.

There are also connections recently discovered between commutative algebra and statistics. Namely, to design a random walk on a lattice is equivalent to constructing a set of generators for the ideal of a variety given implicitly; that is, solving a problem in classical elimination theory. This is of importance in the statistics of medium-sized data sets—for example, contingency tables—where classical methods give wrong answers. The classical methods of elimination theory are hard; the modern technique of Groebner bases is now often used.

Example 2: Interactions Between Mathematics and Theoretical High-Energy Physics

Certainly one of the most important and surprising recent developments in mathematics has been its interaction with theoretical high-energy physics. Large swathes of geometry, representation theory, and topology have been heavily influenced by the interaction of ideas from quantum field theory and string theory, and, in turn, these areas of physics have been informed by advances in the mathematical areas. Examples include the relationship of the Jones polynomial for knots with quantum field theory, and Donaldson invariants for 4-manifolds and the related Seiberg-Witten invariants. Then there is mirror symmetry, discovered by physicists, which, in each original formulation, led to the solution of one of the classical enumerative problems in algebraic geometry, the number of rational curves of a given degree in a quintic hypersurface in projective 4-space. This has been expanded conjecturally to a vast theory relating complex manifolds and symplectic manifolds. A more recent example of the cross-fertilization between mathematics and physics is the reformulation of the geometric Langlands program in terms of quantum field theory. A very recent example involves the computation of scattering amplitudes in gauge theories, motivated by the practical problem of computing backgrounds for the Large Hadron Collider. These computations use the tools of algebraic geometry and some methods from geometric number theory. Ideas of topology are very important in many areas of physics. Most notably, topological quantum field theories of the Chern-Simons form are crucial to understanding some phases of condensed matter systems. These are being actively explored because they offer a promising avenue for constructing quantum computers.

Example 3: Kinetic Theory

Kinetic theory is a good example of the interaction between areas of the mathematical sciences that have traditionally been seen as core and those that had been seen as applied. The theory was proposed by Maxwell and Boltzmann to describe the evolution of rarefied gases (not dense enough to be considered a "flow," not dispersed enough to be just a system of particles, a dynamical system). Mathematically, the Boltzmann equation involves the spatial interaction (collisions) of probability densities of particles travelling at different velocities. The analytical properties of solutions—their existence, regularity, and stability, and the phenomenon of shock formation—were little understood until approximately 30 years ago. Hilbert and Carleman worked on these problems for many years with little success, and attempts to understand the analytical aspects of the equation—existence, regularity, stability of solutions, as well as possible shock formation—had not advanced very far. In the 1980s, the equation arose as part of the need for the modeling of the reentry dynamics of space flight through the upper atmosphere and it was taken up again by the mathematical community, particularly in France. That gave rise to 20 years of remarkable development, from the celebrated work of Di Perna-Lions (1988) showing the existence of solutions, to the recent contributions of Villani and his collaborators. In the meantime, the underlying idea of the modeling of particles interacting at a rarefied scale appeared in many other fields in a more complex way: sticky particles, intelligent particles, and so on, in the modeling of semiconductors, traffic flow, flocking, and social behavior, particularly in phenomena involving decision making.

The sorts of connections exemplified here are powerful, because they establish alternative modes by which mathematical concepts may be explored. They often inspire further work because surprising connections hint at deeper relationships. It is clear that the mathematical sciences have benefited in recent years from valuable, and perhaps surprising, connections within the discipline itself. For example, the Langlands program in geometry is bringing together many different threads of mathematics, such as number theory, Lie theory, and representation groups, and as noted above it has more recently also been linked to physics. Another example is Demetrios Christodoulou's work on the formation of black holes from a few years ago, which resolved a question that had been open for decades through a combination of insights from PDEs and differential geometry.

Fields in the mathematical sciences are mature enough so that researchers know the capabilities and limitations of the tools provided by their field, and they are seeking other tools from other areas. This trend seems to be flourishing, with the result that there is an increase in interdisciplinarity across the mathematical sciences. For example, there is greater interest in combinatorial methods which, 50 years ago, might not have been pursued because those methods may not have elegant structures and because computation may be required. The tendency decades ago was to make simplifying assumptions to eliminate the need for combinatorial calculations. But many problems have a real need for a combinatorial approach, and many researchers today are willing to do those computations. Because of these interdisciplinary opportunities, more researchers are reaching out from areas that might in the past have been self-contained. Also, as discussed below, it is easier to collaborate these days because of the Internet and other communications technologies.

In other research areas, opportunities are created when statistics and mathematics are brought together, in part because the two fields have complementary ways of describing phenomena. An example is found in environmental sciences, where the synergies between deterministic mathematical models and statistics can lead to important insights. Such an approach is helpful for, say, understanding the uncertainties in climate models, because of the value in combining insight about deterministic PDE-based models with statistical insights about the uncertainties.

Because of these exciting opportunities that span multiple fields of the mathematical sciences, the amount of technical background needed by researchers is increasing. Education is never complete today, and in some areas older mathematicians may make more breakthroughs than in the past because so much additional knowledge is needed to work at the frontier. For this reason, postdoctoral research training may in the future become necessary for a greater fraction of students, at least in mathematics. The increase in postdoctoral study has been dramatic over the past 20 years, such that in the fall of 2010, 40 percent of recent Ph.D. recipients in the mathematical sciences were employed as postdoctoral researchers.[1] Thus, the time from receipt of the Ph.D. to attaining a tenure-track academic

---

[1] R. Cleary, J.W. Maxwell, and C. Rose, 2010, Report on the 2009-2010 new doctoral recipients. *Notices of the AMS* 58(7):944-954.

position has lengthened. While these trends clearly create researchers with stronger backgrounds, the length of time for becoming established as a researcher could lessen the attractiveness of this career path.

At the same time, more mathematical scientists are now addressing applications, such as those in computer science. This work builds successfully on the deep foundations that have been constructed within mathematics. For example, results of importance to computer science have been achieved by individuals who are grounded in discrete mathematics and combinatorics and who may not have had previous exposure to the particular application.

These increasing opportunities for interdisciplinary research pose some challenges for individuals and the community. Interdisciplinary work is facilitated by proximity, and even a walk from one corridor to another can be a hindrance. So attention must be paid to fostering collaboration, even within a single department. When the connections are to be established across disciplines, this need is even more obvious. Ideally, mathematical scientists working in biology, for example, will spend some of their time visiting experimental labs, as will mathematical scientists working with other disciplines. But to make this happen, improved mechanisms for connecting mathematical scientists with potential collaborators are needed, such as research programs that bring mathematical scientists and collaborators together in joint groups. Such collaborations work best when the entire team shares one primary goal—such as addressing a question from biology—even for the team members who are not biologists per se. But to make this work we need adjustments to reward systems, especially for the mathematical scientists on such teams.

One leading researcher who spoke with the committee observed how mathematicians at Microsoft Research are often approached by people from applied groups, which is a fortunate result of the internal culture. One value that mathematicians can provide in such situations that is often underestimated is that they can prove negative results: that is, the impossibility of a particular approach. That knowledge can redirect a group's efforts by, say, helping them realize they should attack only a limited version of their problem or stopping them from expending more resources on a hopeless task. This ultimately increases productivity because it helps the organization to focus resources better. This is a contribution of mathematics beyond product-focused work or algorithm development.

Another set of challenges to interdisciplinary students and researchers stems from their lack of an obvious academic home. Who is in their community of peers? Who judges their contributions? How are research proposals and journal submissions evaluated? At the National Institutes of Health, for example, the inclusion of mathematical scientists in the study sections that review proposals with mathematical or statistical content is an important step, though it is not a perfect process. Tenure review in large universities can also be problematic for interdisciplinary junior faculty. While it takes longer to build a base of interdisciplinary knowledge, once that base is built it can open doors to very productive research directions that are not feasible for someone with a more conventional background. Universities are changing slowly to recognize that interdisciplinary faculty members can produce both better research and better education. There is a career niche for such people, but it could be improved. For example, it might be necessary to relax the tenure clock for researchers pursuing interdisciplinary topics, and a proper structure must be in place in order to conduct appropriate tenure reviews for them. This is one way to break down academic silos.

INNOVATION IN MODES FOR SCHOLARLY INTERACTIONS AND PROFESSIONAL GROWTH

Mathematical Science Institutes

A major change in the mathematical sciences over the past decade and more has been the increasing number of mathematical science institutes and their increased influence on the discipline and community. In 1981 there was only one such institute in the United States—the Institute for Advanced Study in Princeton, which has a very different character from the institutes created after it. The National Science Foundation/Division of Mathematical Sciences (NSF/DMS) now supports eight mathematical sciences institutes in the United States;[2] other entities heavily involved in the mathematical sciences include the Clay Institute for Mathematics, the Simons Center for Geometry and Physics, and the Kavli Institute for Theoretical Physics. In the last 20 years we have seen new institutes appear in Japan, England, Ireland, Canada, and Mexico, to name just a few countries, joining older institutes in France, Germany, and Brazil. Overall, there are now some 50 mathematical science institutes in 24 different countries.[3] These institutes have made it easier for mathematical scientists to form and work in collaborative teams that bridge two or more fields or that connect the mathematical sciences to another discipline.

Typical goals of most U.S. mathematical science institutes include the following:

- Stimulate research, collaboration, and communication;
- Seed and sustain important research directions;
- Promote interdisciplinary research;

---

[2] Representing 12.5 percent ($32.5 million) of DMS's 2012 budget request.
[3] *Notices of the AMS*, August 2011, pp. 973-976.

- Build research teams, including collaborations with industry, government; laboratories, and international colleagues;
- Enrich and invigorate mathematics education at all levels;
- Provide postdoctoral training; and
- Expand mathematics opportunities for underrepresented groups.[4]

The institutes promote research and collaboration in emerging areas, encourage continued work on important problems, tackle large research agendas that are outside the scope of individual researchers, and help to maintain the pipeline of qualified researchers for the future. Many of the institute programs help researchers broaden their expertise, addressing the need for linking multiple fields that was emphasized above and in Chapter 3. For example, every year the Institute for Mathematics and its Applications (IMA) offers intensive two-week courses aimed at helping to introduce established researchers to new areas; recent courses have focused on mathematical neuroscience, economics and finance, applied algebraic topology, and so on. All of these institutes have visiting programs, often around a specific theme that varies from year to year, and they invite mathematical scientists from around the world to visit and participate in the programs. This has led to an enormous increase in the cross-fertilization of ideas as people from different places and different disciplines meet and exchange ideas. In addition, it is quite common for these institutes to record the lectures and make them freely available for downloading. Real-time streaming of lectures is just starting to emerge. All of these steps help to strengthen the cohesiveness of the community.

Beyond this, the institutes frequently allow researchers to meet people they would not otherwise meet. This is especially crucial in connecting researchers from other disciplines with the right mathematical scientists. Often, scientists, engineers and medical researchers do not know what mathematics and statistics are available that might be relevant to their problem, and they do not know whom to turn to. Likewise, mathematical scientists are often sitting on expertise that would be just what is needed to solve an outside problem, but they are unaware of the existence of these problems or of who might possess the relevant data.

Arguably, the institutes have collectively been one of the most important vehicles for culture change in the mathematical sciences. Some illustrations of the impact of mathematical science institutes follow.[5]

To help the mathematical sciences build connections, the IMA reaches out so that some 40 percent of the participants in its programs come from

---

[4] These goals are explicit for the NSF-supported institutes.

[5] The committee thanks IMA director Fadil Santosa, IPAM director Russel Caflisch, SAMSI director Richard Smith, and MSRI director Robert Bryant for helpful inputs to this section.

outside the mathematical sciences. In this way, it has helped to nucleate new communities and networks in topics such as mathematical materials science, applied algebraic geometry, algebraic statistics, and topological methods in proteomics. The Institute for Pure and Applied Mathematics (IPAM) hosts a similar percentage of researchers from other disciplines.

The institutes have had success in initiating new areas of research. For example, IPAM worked for 9 years to nucleate and then nurture a new focused area of privacy research, starting with a workshop on contemporary methods in cryptography in 2002. That led to a 2010 workshop on statistical and learning-theoretic challenges in data privacy, which brought together data privacy and cryptography researchers to develop an approach to data privacy that is motivated and informed by developments in cryptography, one of them being mathematically rigorous concepts of data security. A second follow-on activity was a 2011 workshop on mathematics of information-theoretic cryptography, which saw algebraic geometers and computer scientists working on new approaches to cryptography based on the difficulty of compromising a large number of nodes on a network. Another IPAM example illustrates that the same process can be important and effective in building connections within the discipline. A topic called "expander graphs" builds on connections that have emerged among discrete subgroups of Lie groups, automorphic forms, and arithmetic on the one hand, and questions in discrete mathematics, combinatorics, and graph theory on the other. In 2004 IPAM held the workshop Automorphic Forms, Group Theory, and Graph Expansion, which was followed by a program on the subject at the Institute for Advanced Study in 2005 and a second IPAM workshop, Expanders in Pure and Applied Mathematics, in 2008. Similarly, the Statistical and Applied Mathematical Sciences Institute (SAMSI) has worked to develop the general topic of low-dimensional structure in a high-dimensional system. Many problems of modern statistics arise when the number of available variables is comparable with or larger than the number of independent data points (often referred to as the $p > n$ problem). Traditional methods for dealing with such problems involve techniques such as variable selection, ridge regression, and principal components regression. Beginning in the 1990s, more modern methods such as lasso regression and wavelet thresholding were developed. These ideas have now been extended in numerous directions and have attracted the attention of researchers in computer science, applied mathematics, and statistics, in areas such as manifold learning, sparse modeling, and the detection of geometric structure. This is an area with great potential for interaction among statisticians, applied mathematicians, and computer scientists.

The Mathematical Sciences Research Institute (MSRI) is focused primarily on the development of fundamental mathematics, specifically in areas in which mathematical thinking can be applied in new ways. Programs have spanned topics such as mathematical biology, theoretical and applied topology, mathematics of visual analysis, analytic and computational elliptic and parabolic equations, dynamical systems, geometric evolution equations, parallel computing for mathematics, computational finance, statistical computing, multiscale methods, climate change modeling, algebraic geometry, and advances in algebra and geometry.

In spite of its primary focus on the mathematical sciences per se, MSRI has long included a robust set of outreach activities. For example, its 2006 program Computational Aspects of Algebraic Topology explored ways in which the techniques of algebraic topology are being applied in various contexts related to data analysis, object recognition, discrete and computational geometry, combinatorics, algorithms, and distributed computing. That program included a workshop focused on application of topology in science and engineering, which brought together people working in problems ranging from protein docking, robotics, high-dimensional data sets, and sensor networks. In 2007, MSRI organized and sponsored the World Congress on Computational Finance, in London, which brought together both theoreticians and practitioners in the field to discuss its current problems. MSRI has also sponsored a series of colloquia to acquaint mathematicians with fundamental problems in biology. An example is the 2009 workshop jointly sponsored by MSRI and the Jackson Laboratories on the topic of mathematical genomics. Both MSRI and SAMSI have helped build bridges between statisticians and climate scientists through at least six programs focused on topics such as new methods of spatial statistics for climate change applications, data assimilation, analysis of climate models as computer experiments, chaotic dynamics, and statistical methods for combining ensembles of climate models.

IPAM and SAMSI provide two additional illustrations of how the institutes build new connections to other disciplines. After a professor of Scandinavian languages at UCLA participated in a 2007 IPAM program on knowledge and search engines, which introduced him to researchers and methods from modern information theory, he went on to organize two workshops in 2010 and 2011, on networks and network analysis for the humanities; the workshops were funded by the National Endowment for the Humanities and cosponsored by IPAM. They led to the exploration of new data analysis tools by many of the humanists who participated. SAMSI's example comes from a more established area, the interaction between statistics and social sciences, which SAMSI has supported through several activities, such as a workshop to explore computational methods for causal modeling and for the analysis of transactions and social relationships.

The IMA has a long history of outreach to industry, for instance through its Industrial Postdoctoral Fellowship program and other activities that bring cutting-edge mathematical sciences to bear on significant industrial problems. Examples include uncertainty quantification in the automotive industry and numerical simulation of ablation surgery. Overall, IMA has trained over 300 postdoctoral fellows since 1982, and about 80 percent of them are now in academic positions. The IMA also offers programs for graduate students, most notably its regular workshops on mathematical modeling in industry, in which students work in teams under the guidance of industry mentors on real-world problems from their workplace. Through this program, many mathematical scientists have been exposed early in their careers to industrial problems and settings.

As an example of value provided beyond academe, IPAM has been instrumental in introducing modern imaging methods to the National Geospatial-Intelligence Agency (NGA). Several individuals from NGA attended IPAM's 2005 summer school, Intelligent Extraction of Information from Graphs and High-Dimensional Data, which convinced them and their agency to explore further. Subsequently, NGA organized a series of three workshops at IPAM on advancing the automation of image analysis. This led to hiring by NGA of several new mathematics Ph.D.s with expertise in image analysis, as well as a major funding initiative. IPAM has similarly held workshops for the Office of Naval Research (ONR) on aspects of machine reasoning, which may lead to an ONR funding initiative. IPAM's program Multiscale Geometry and Analysis in High Dimensions led to the explosive growth of applications of compressed sensing, followed by a large funding program at DARPA.

In addition to the institutes, NSF/DMS and other financial supporters of mathematics have in recent years created other funding programs to encourage and nurture research groups, which help investigators to address broad and cross-cutting topics.

Changing Models for Scholarly Communication

The first thing that comes to mind when one thinks of interconnectivity these days is the Internet and the World Wide Web. These affect practically all human activity, including the way that mathematical scientists work. The maturation of the Internet has led in the past 15-20 years to the availability of convenient software tools that painlessly lead to the quick dissemination of research results (consider for instance the widely used arXiv preprint server, http://arxiv.org/), the sharing of informal ideas through blogs and other venues, and the retrieval of information through efficient search engines. These new tools have profoundly changed both the modes of collaboration and the ease with which mathematical scientists can work across fields. The existence of arXiv has had a major influence on scholarly communication in the mathematical sciences, and it will probably become even more important. The growth of such sites has already had a great impact on the traditional business model for scientific publishing, in all fields. It is difficult to say what mode(s) of dissemination will predominate in 2025, but the situation will certainly be different from that of today.

The widespread availability of preprints and reprints online has had a tremendous democratizing influence on the mathematical sciences. Gone are the days where you had to be in Paris to be among the first to learn about Serre or Grothendieck or Deligne's latest ideas. Face-to-face meetings between mathematical scientists remain an essential mode of communication, but the tyranny of geography has substantially lessened its sway. However, the committee is concerned about preserving the long-term accessibility of the results of mathematical research. Rapid changes in the publishing industry and the fluidity of the Internet are also of concern. This is a very uncertain time for traditional scholarly publishing,[6] which in turn raises fundamental concerns about how to share and preserve research results and maintain assured quality. Public archives such as arXiv play a valuable role, but their long-term financial viability is far from assured, and they are not used as universally as they might be. The mathematical sciences community as a whole, through its professional organizations, needs to formulate a strategy for maximizing public availability and the long-term stewardship of research results. The NSF could take the lead in catalyzing and supporting this effort.

Thanks to mature Internet technologies, it now is easy for mathematical scientists to collaborate with researchers across the world. One striking instance of this globalization of the mathematical sciences is the first "polymath" projects, which were launched in 2009.[7] To quote from Terence Tao, these "are massively collaborative mathematical research projects, completely open for any interested mathematician to drop in, make some observations on the problem at hand, and discuss them with the other participants."[8] Another recent phenomenon is global review of emerging ideas. Not only do such projects contribute to advancing research, but they also serve to locate other researchers with the same interest and with the right kind of expertise; they represent an ideal vehicle for expanding personal collaborative networks. However, new modes of collaboration and "publishing" will call for adjustments in the methods for quality control and for rewarding professional accomplishments.

---

[6] See, for example, Thomas Lin, Cracking open the scientific process. *The New York Times*. January 17, 2012.

[7] See Timothy Gowers and Michael Nielsen, 2009, Massively collaborative mathematics. *Nature* 461: 879-881.

[8] Quoted from http://terrytao.wordpress.com/2009/09/17/a-speech-for-the-american-academy-of-arts-and-sciences. Accessed March 19, 2012.

Widespread dissemination of research results has made it easier for anyone to borrow ideas from other fields, thereby creating new bridges between subdisciplines of the mathematical sciences or between the mathematical sciences and other fields of science, engineering, and medicine. For example, new research directions can be seen where ideas from abstract probability theory prove to have very deep consequences in signal processing and tantalizing applications in signal acquisition, and where tools from high-dimensional geometry can change the way we perform fundamental calculations, such as solving systems of linear equations. Effortless access to information has spurred the development of communities with astonishingly broad collective expertise, and this access has lowered barriers between fields. In this way, theoretical tools find new applications, while applications renew theoretical research by offering new problems and suggesting new directions. This cycle is extremely healthy.

A recent paper[9] evaluated an apparent shift in collaborative behavior within the mathematical sciences in the mid-1990s. At that time, the networks of researchers in core and applied mathematics moved from being centered primarily around a small number of highly prolific authors toward networks displaying more localized connectivity. More and stronger collaboration was in evidence. Brunson and his collaborators speculated that a cause of this trend was the rise of e-communications and the Web—for example, arXiv went online in 1993 and MathSciNet in 1996—because applied subdisciplines, which historically had made greater use of computing resources, showed the trend most strongly.

The Internet provides a ready mechanism for innovation in communication and partnering, and novel mechanisms are likely to continue to appear. As just one more example, consider the crowdsourcing, problem-solving venture called InnoCentive. It is an example of another new Web-enabled technology that may have real impacts on the mathematical sciences by providing opportunities to learn directly of applied challenges from other disciplines and to work on them. InnoCentive is backed by venture capitalists with the goal of using "crowdsourcing"—Web-based methods for parceling out tasks to anyone who wishes to invest time in hopes of achieving results and then receiving payment—to solve problems for corporate, government, and nonprofit clients. When checked on March 16, 2012, the company's Web site listed 128 challenges that were either open or for which submissions were being evaluated. Of these, 13 were flagged as having mathematical or statistical content. Examples of the latter included challenges such as the development of an algorithm to identify underlying geometric features

---

[9] Brunson, J.C., S. Fassino, A. McInnes, M. Narayan, B. Richardson, C. Franck, P. Ion, and R. Laubenbacker, 2012, Evolutionary Events in a Mathematical Sciences Research Collaboration Network. Manuscript submitted for publication. arXiv:1203.5158 [physics.soc-ph].

in noisy two-dimensional data and the creation of a model to predict particle size distributions after milling. The fee for supplying a best solution to these problems typically ranged from $15,000 to $30,000, and most challenges attracted hundreds of potential solvers.

There are debates about whether crowdsourcing is a healthy trend for a research community. And while InnoCentive does offer opportunities for mathematical scientists to engage in a broad range of problems, the engagement is at arm's length, at least initially. However, crowdsourcing is one more Web-based innovation that may affect mathematical scientists, and the community should be aware of it.

THE MATHEMATICAL SCIENCES SHOULD MORE THOROUGHLY EMBRACE COMPUTING

Computing is often the means by which the mathematical sciences are applied to other fields. For example, mathematical scientists collaborate with astrophysicists, neuroscientists, or materials scientists to develop new models and their computationally feasible instantiations in software in order to simulate complex phenomenology. The mathematical sciences can obviously contribute to creating mathematical and statistical models. But they can also contribute to the steps that translate those models into computational simulations: discretizations, middleware (such as gridding algorithms), numerical methods, visualization methods, and computational underpinnings.

Whenever a computer simulation is used, the question of validation is critical. Validation is an essential part of simulation, and the mathematical sciences provide an underlying framework for validation. Because of that, validation is also an exciting growth area for the mathematical sciences. Owing to the multidisciplinary nature of validation studies, mathematical scientists may increasingly engage with behavioral scientists, domain scientists, and risk and decision analysts along this growing frontier.

This spectrum of challenges is often labeled "scientific computing," an area of study in its own right and an essential underpinning for simulation-based engineering and science. It is, however, somewhat of an academic orphan: that is, a discipline that does not fit readily into any one academic department and one for which academic reward systems do not align well with the work that needs to be carried out. Scientific computing experts must know enough about the domain of the work to ensure that the software is properly approximating reality, while also understanding relevant applied mathematics and the subtleties of a computer's architecture and compiler. But those disciplines do not normally assign a high priority to the combination of skills possessed by the scientific computing expert, nor is the development of critical and unique software rewarded as an academic achievement. At different academic institutions, scientific computing might at times be found within departments of computer science or applied mathematics or at other times be spread across a range of science and engineering departments. In the latter case, scientific computing experts often are faced with the misalignment of rewards and incentives.

Recognizing that many aspects of scientific computing are at heart very mathematical, departments in the mathematical sciences should play a role in seeing that there is a central home for scientific computing research and education at their institutions, whether or not it is within their own departments. Computation is central to the future of the mathematical sciences, and to future training in the mathematical sciences.

Another aspect of computing vis-á-vis the mathematical sciences is the availability of more and better data, including computer-generated data. There is a long and strong tradition in mathematics to discount empirical evidence; while occasionally someone will refer to empirical observations as having suggested a research direction, even that is rare. But today, with advanced computing, we have the opportunity to generate a great deal of empirical evidence, and this trend is growing. However, some mathematical scientists will be deterred from seizing this opportunity unless we overcome the tradition and embrace phenomenology-driven research in addition to the theorem/proof paradigm. Computational resources open the door to more data-driven mathematical exploration.

More generally, because computation is often the means by which methods from the mathematical sciences are applied in other disciplines and is also the driver of many new applications of the mathematical sciences, it is important that most mathematical scientists have a basic understanding of scientific computing. Academic departments may consider seminars or other ways to make it easy for mathematical scientists to learn about and keep up with the rapidly evolving frontiers of computation.

Some mathematical sciences research would benefit from the most advanced computing resources, and not many mathematical scientists are currently exploiting those capabilities. Because the nature and scope of computation is continually changing, there is a need for a mechanism to ensure that mathematical sciences researchers have access to computing power at an appropriate scale. NSF/DMS should consider instituting programs to ensure that state-of-the art computing power is widely available to mathematical science researchers.

FUNDING IMPLICATIONS OF INCREASING CONNECTIVITY OF THE MATHEMATICAL SCIENCES

The broadening of the mathematical sciences discussed in Chapter 3 raises the question of whether the mathematical science enterprise is expanding along with its opportunities. This is very difficult to answer, but the committee is concerned that the accelerating reach of the mathematical sciences is not being matched by a commensurate increase in financial support. A key message of the 1984 David report (see Appendix A) was that, in order to maintain a pipeline of students to replenish the research capabilities in place *at that time*, federal funding for mathematical sciences research would need to double. That doubling, in inflation-adjusted dollars, finally occurred late in the 1990s, as shown in Table 4-1. However, that doubling goal was based on the David report's estimate of what it would take to adequately fund the 2,600 mathematical scientists who were actively performing research in 1984. In light of the dramatic expansion of research opportunities described in Chapter 3 and the commensurately larger pipeline of students requiring mathematical science education, the David report's goals are clearly not high enough to meet today's needs. While the funding for NSF/DMS did reach the goals of the David report, and later exceeded them, the overall growth in federal research funding for the mathematical sciences has not been on the same scale as the growth in intellectual scope documented in this report. The recent addition of a large private source, through the Simons Foundation, while very welcome, can only be stretched so far.

As shown in Table 4-1, Department of Defense (DOD) funding for the mathematical sciences has increased only about 50 percent in constant dollars since 1989. It is difficult to discern trends at the Department of Energy (DOE) or the National Institutes of Health (NIH), and the "other agency" category remains oddly underfunded given the pervasiveness of the mathematical sciences. Apropos that last point, some science agencies—such as the National Oceanic and Atmospheric Administration, the National Aeronautics and Space Administration, the Environmental Protection Agency, and the U.S. Geological Survey at present have few direct interfaces with the mathematical sciences community and provide very little research support for that community. And although many agencies that deal with national security, intelligence, and financial regulation rely on sophisticated computer simulations and complex data analyses, only a fraction of them are closely connected with mathematical scientists. As a result, the mathematical sciences are unable to contribute optimally to the full range of needs within these agencies, and the discipline—especially its core areas—is as a result overly dependent on NSF.

> **Conclusion 4-1: The dramatic expansion in the role of the mathematical sciences over the past 15 years has not been matched by a comparable expansion in federal funding, either in the total amount or in the diversity of sources. The discipline—especially the core areas—is still heavily dependent on the NSF.**

TABLE 4-1 Federal Funding for the Mathematical Sciences (in millions)

	1984 (Constant 2011 Dollars)	1989 (Constant 2011 Dollars)	1998 (Constant 2011 Dollars)	2010 Estimate (Constant 2011 Dollars)	2012 Estimate (Actual Dollars)
NSF					
DMS	113	143	147	251	238
Other	11	17	—	—	—
Subtotal	124	160	147	251	238
DOD					
AFOSR	28	36	48	53	47
ARO	19	25	19	12	16
DARPA	—	19	31	12	28
NSA	—	6	—	7	6
ONR	33	28	12	20	24
Subtotal	80	113	111	105	121
DOE					
Applied mathematics	—	—	—	45	46
SciDAC	—	—	—	51	44
MICS	—	—	194	—	—
Subtotal	8	14	194	96	90
NIH					
NIGMS	—	—	—	51	—
NIBIB	—	—	—	40	—
Subtotal	—	—	—	91	—
Other agencies	6	3	—	—	—
Total	217	289	451	543	449

NOTE: Inflation calculated using the higher education price index (HEPI) outlined in Common Fund Institute, "2011 Update: Higher Education Price Index," Table A. Available at http://www.commonfund.org/CommonFundInstitute/HEPI/HEPI%20Documents/2011/CF_HEPI_2011_FINAL.pdf. Acronyms: AFOSR, Air Force Office of Scientific Research; ARO, Army Research Office; DARPA, Defense Advanced Research Projects Agency; MICS, Mathematical, Information, and Computational Sciences Division; NIBIB, National Institute of Biomedical Imaging and Bioengineering; NIGMS, National Institute of General Medical Sciences; NSA, National Security Agency; ONR, Office of Naval Research; SciDAC, Scientific Discovery Through Advanced Computing.
SOURCES: 1984 and 1989 data from David II; 1998 data from Daniel Goroff, 1999, Mathematical sciences in the FY 2000 budget, *Notices of the AMS* 46(6):680-682; 2010 and 2012 data from Samuel Rankin, III, 2011, Mathematical sciences in the FY 2012 budget, *AAAS Report XXXVI: Research and Development FY 2012*:225-230.

In 2010, NSF/DMS was externally reviewed by a Committee of Visitors (COV) to address its balance, priorities, and future directions, among other things. This COV review culminated in a report that found that DMS is underfunded and that, in spite of an overall budget increase, most of the increased funds went to interdisciplinary programs while funding of the core DMS programs stayed constant. The COV recommended that more money be directed to core areas.[10] In its response to the COV report, DMS pointed out that funding for core areas increased significantly from 2006 to 2007, with additional (but small) increases in the following 2 years.[11] DMS is faced with an innate conflict: As the primary funding unit charged with maintaining the health of the mathematical sciences, it is naturally driven to aid the expansions discussed in Chapter 3; yet it is also the largest of a very few sources whose mission includes supporting the foundations of the discipline, and thus it plays an essential role with respect to those foundations. As noted in Chapter 3, some mathematical scientists receive research support from other parts of NSF and from nonmath units in other federal funding agencies, but there are only anecdotal accounts of this. With limited data it is difficult to get a full picture of the totality of funding for the broader mathematical sciences community—the community that is an intellectually coherent superset of those researchers who sit in departments of mathematics or statistics—and to determine whether the funding is adequate and appropriately balanced. Nor can we say whether it is keeping pace with the expanding needs of this broader community. There are challenges inherent in supporting a broad, loosely knit community while maintaining its coherence, and the adequacy and balance of funding is a foremost concern. As noted in Chapter 3, funding of excellence wherever it is found should still be the top priority.

A VISION FOR 2025

Finding: Mathematical sciences work is becoming an increasingly integral and essential component of a growing array of areas of investigation in biology, medicine, social sciences, business, advanced design, climate, finance, advanced materials, and much more. This work involves the integration of mathematics, statistics, and computation in the broadest sense, and the interplay of these areas with areas of potential application; the mathematical sciences are best conceived of as including

---

[10] NSF/DMS Committee of Visitors, NSF/DMS, 2010, *Report of the 2010 Committee of Visitors*. Available at http://www.nsf.gov/attachments/117068/public/DMS_COV_2010_final_report.pdf.

[11] NSF, 2010, *Response to the Division of Mathematical Sciences Committee of Visitors Report*. Available at http://www.nsf.gov/mps/advisory/covdocs/DMSResponse_2010.pdf.

all these components. These activities are crucial to economic growth, national competitiveness, and national security. This finding has ramifications for both the nature and scale of funding of the mathematical sciences and for education in the mathematical sciences.

Conclusion 4-2: The mathematical sciences have an exciting opportunity to solidify their role as a linchpin of twenty-first century research and technology while maintaining the strength of the core, which is a vital element of the mathematical sciences ecosystem and essential to its future. The enterprise is qualitatively different from the one that prevailed during the latter half of the twentieth century, and a different model is emerging—one of a discipline with a much broader reach and greater potential impact. The community is achieving great success within this emerging model, as recounted in this report. But the value of the mathematical sciences to the overall science and engineering enterprise and to the nation would be heightened by increasing the number of mathematical scientists who share the following characteristics:

- They are knowledgeable across a broad range of the discipline, beyond their own area(s) of expertise;
- They communicate well with researchers in other disciplines;
- They understand the role of the mathematical sciences in the wider world of science, engineering, medicine, defense, and business; and
- They have some experience with computation.

It is by no means necessary or even desirable for all mathematical scientists to exhibit these characteristics, but the community should work toward increasing the fraction that does.

To move in these directions, the following will need attention:

- The culture within the mathematical sciences should evolve to encourage development of the characteristics listed in Conclusion 4-2.
- The education of future generations of mathematical scientists, and of all who take mathematical sciences coursework as part of their preparation for science, engineering, and teaching careers, should be reassessed in light of the emerging interplay between the mathematical sciences and many other disciplines.
- Institutions, for example, the funding mechanisms and reward systems—should be adjusted to enable cross-disciplinary careers when they are appropriate.
- Expectations and reward systems in academic mathematics and statistics departments should be adjusted so as to encourage a

broad view of the mathematical sciences and to reward high-quality work in any of its areas.

- Mechanisms should be created that help connect researchers outside the mathematical sciences with mathematical scientists who could be appropriate collaborators. Funding agencies and academic departments in the mathematical sciences could play a role in lowering the barriers between researchers and brokering such connections. For academic departments, joint seminars, cross-listing of courses, cross-disciplinary postdoctoral positions, collaboration with other departments in planning courses, and courtesy appointments are useful in moving this process forward.
- Mathematical scientists should be included more often on the panels that design and award interdisciplinary grant programs. Because so much of today's science and engineering builds on advances in the mathematical sciences, the success and even the validity of many projects depends on the early involvement of mathematical scientists.
- Funding for research in the mathematical sciences must keep pace with the opportunities.

While there are limits to the influence that it can have on the direction and character of research in the mathematical sciences and on the culture of the mathematical sciences community, the NSF can exercise leadership and serve as an enabler of positive developments. Successful examples include the flourishing Research Experiences for Undergraduates program and NSF's portfolio of mathematical science institutes. The NSF can, through funding opportunities, enhance the pace of change and facilitate bottom-up developments that capitalize on the energy of members of the community—examples include open calls for workforce proposals, grants to enable the development of new courses and curricula; grants that support interdisciplinary research and research between disciplines within the mathematical sciences, grants that enable individuals to acquire new expertise; and programs that make it easier for young people to acquire experience in industry and to acquire international experience.

The trends discussed in this chapter may appear quite disruptive to many core mathematicians, or even irrelevant. To address that possibility, the committee closes with a personal reflection by the study vice-chair, Mark Green, in Box 4-1, "Core Mathematicians and the Emerging Mathematical Landscape."

BOX 4-1
Core Mathematicians and the Emerging
Mathematical Landscape

by Mark Green

Al Thaler, a long-time program officer at the NSF, once told me, "The twenty-first century is going to be a playground for mathematicians." Events have more than justified his prediction. Core mathematics is flourishing, new ways of using the mathematical sciences continue to develop, more and more disciplines view the mathematical sciences as essential, and new areas of mathematics and statistics are emerging. Where do core mathematicians fit into all this?

Of course, their main role in research is to continue to produce excellent core mathematics. That said, for those who are interested, the intellectual challenges coming from other fields, and the opportunities they present for core mathematicians are extraordinary. Core mathematicians often have a knowledge base and a set of insights and instincts that would be of immense value in other fields.

As someone who was trained in the late 1960s and early 1970s at two outstanding U.S. mathematics departments, the things I was exposed to reflected the emphasis of that time. I had no classes—even in the lower division—that dealt with probability or statistics. Although it was offered, I took no discrete mathematics. The only algorithm I saw at either place was Gaussian elimination, and the last algorithm before that was long division. Throughout much of my career I didn't experience these as gaping lacunae, but it is definitely not how I would train a student these days.

When I became the director of an NSF mathematical science institute, one of the first things I started doing was reading *Science* and *Nature*, two places where scientists publish work they feel will be of interest to the broader scientific community. What you see immediately is that almost none of the articles are specifically about mathematics, but that the majority make use of sophisticated mathematical and statistical techniques. This is not surprising—when a new idea in core mathematics first sees the light of day, researchers in other fields do not know how to make use of it, and by the time the new idea has made its way into wider use, it is no longer new. The overwhelming impression one comes away with from reading these journals is what an explosively creative golden age of science we are living through and just how central a role the mathematical sciences are playing in making this possible.

Examples abound. An extensive and very beautiful literature developed on Erdős-Rényi random graphs—these are graphs where a fraction p of the $n(n-1)/2$ possible edges of a graph with n vertices is filled in at random, each edge being equally likely. However, as very large naturally occurring graphs—citations and collaborations, social networks, protein interaction networks, the World Wide Web—became widely available for study, it became apparent that they did not look at all like Erdős-Rényi random graphs. It is still hotly contested what class or classes of probabilistically generated graphs best describe those that actually occur. This is the embryonic stage of an emerging area of study, with questions about how best

continued

BOX 4-1 Continued

to design a network so that it has certain properties, how best to ascertain the structure of a network by the things one can actually measure, and how to search the network most effectively. There are purely mathematical questions relating to how to characterize asymptotically different classes of large random graphs and, for such classes, potential analogs of the theorems about Erdős-Rényi graphs and the cutoff value of p for which there is an infinite connected component of the graph.

With the advent of digital images, the question of how to analyze them—to get rid of noise and blurring, to segment them into meaningful pieces, to figure out what objects they contain, to recognize both specific classes of objects such as faces and to identify individual people or places—poses remarkably interesting mathematical and statistical problems. Core mathematicians are aware of the extraordinary work of Fields medalist David Mumford in algebraic geometry, but many may be unaware of his seminal work in image segmentation (the Mumford-Shah algorithm, for example). Approaches using a moving contour often involve geometrically driven motion—for example, motion by curvature—and techniques such as Osher-Vese based in analysis involve decompositions of the image intensity function into two components, one minimizing total variation (this piece should provide the "cartoon") and one minimizing the norm in the dual of the space of functions of bounded variation (this piece should provide the "texture").

In machine learning, the starting point for many algorithms is finding a meaningful notion of distance between data points. In some cases, a natural distance suggests itself—for example, the edit distance for comparing two sequences of nucleotides in DNA that appear in different species where the expected relationship is by random mutation. In other cases, considerable insight is called for—to compare two brain scans, one needs to "warp" one into the other, requiring a distance on the space of diffeomorphisms, and here there are many interesting candidates. For large data sets, the distance is sometimes found using the data set itself—this underlies the method of diffusion geometry, which relates the distance between two data points to Brownian motion on the data set, where only a very local notion of distance is needed to get started. There are interesting theoretical problems about how various distances can be bounded in terms of one another, and to what extent projections from a high-dimensional Euclidean space to a lower-dimensional one preserves distances up to a bounded constant. This is one facet of dimensionality reduction, where one looks for lower-dimensional structures on which the dataset might lie.

Many of these problems are part of large and very general issues—dealing with "big data," understanding complex adaptive systems, and search and knowledge extraction, to name a few. In some cases, these represent new areas of mathematics and statistics that are in the process of being created and where the outlines of an emerging field can only be glimpsed "through a glass, darkly." Research in core mathematics has a long track record of bringing the key issues in an applied problem into focus, finding the general core ideas needed, and thereby enabling significant forward leaps in applications. We take this for granted when

continued

BOX 4-1 Continued

we look at previous centuries, but the same phenomenon is providing opportunities and challenges today.

There is a long list of ways mathematics is now being used, and the types of fundamental mathematics that are needed spans almost every field of core mathematics—algebra, geometry, analysis, combinatorics, logic. A sampling of these uses, described in nontechnical language, can be found in the companion volume to this report, *Fueling Innovation and Discovery: The Mathematical Sciences in the 21st Century*.

Whether or not one gets directly involved in these developments, it would be very useful to the profession if core mathematicians were to increase their level of awareness of what is going on out there. As educators, professors want to continue to instill in their students the clarity and rigor that characterizes core mathematics. But they must do this cognizant of the fact that what students need to learn has vastly expanded, and multiple educational paths must be available to them. There is, of course, some intellectual investment for core mathematicians involved in teaching these courses, but in the end they will have a wider and more varied range of choices of what to teach, and they will be enlarging the number of ecological niches available to mathematicians in both the academic world and the outside world.

5

People in the
Mathematical Sciences Enterprise

INTRODUCTION

The growth and broadening of research opportunities described in Chapters 3 and 4 necessitate changes in the way students are prepared, along with planning about how to attract a sufficient number of talented young people into the discipline. From its discussions with representatives from industry and government who hire mathematically trained individuals, plus other information cited in Chapters 3-4, the committee concludes that demand for people with strong mathematical science skills is already growing and will probably grow even more. The range of positions that require mathematical skills is also expanding, as more and more fields are presented with the challenges and opportunities of large-scale data analysis and mathematical modeling. While these positions can be filled by individuals with a variety of postsecondary degrees, all of them will need strong skills in the mathematical sciences. This has implications for the mathematical sciences community in its role as educators with a responsibility to prepare students from many disciplines to be ready for a broad range of science, technology, engineering, and mathematics (STEM) careers. Indeed, producing an adequate number of people with expertise in the mathematical sciences at the bachelor's, master's, and Ph.D. levels and an adequate pipeline of well-trained students emerging from grades K-12 is necessary if the STEM fields are to thrive.

CHANGING DEMAND FOR THE MATHEMATICAL SCIENCES

At its meeting in December 2010, the committee heard from four people in sectors of industry that are becoming increasingly reliant on the mathematical sciences:

- Nafees Bin Zafar, who heads the research division at DreamWorks Animation,
- Brenda Dietrich, vice president for Business Analytics and Mathematical Sciences at IBM's T.J. Watson Research Center,
- Harry (Heung-Yeung) Shum, head of Core Search Development at Microsoft Research, and
- James Simons, head of Renaissance Technologies.

The goal of this discussion was to gain insight about some of the topical areas in which the mathematical sciences are critical. Speakers brought knowledge of the demand for the mathematical sciences in the financial sector and the growing demands in business analytics, the entertainment sector, and the information industry. Because Dr. Shum had been a founder of Microsoft Research–Asia, in Beijing, he was also able to comment on the growing mathematical science capabilities in China. The focus of these interactions was to learn about current and emerging uses of mathematical sciences skills, whether or not carried out by people who consider themselves to be mathematical scientists. The financial sector, for example, employs thousands of financial engineers, only a fraction of whom have a terminal degree in mathematics or statistics. (Many are trained in physics or economics, and many have M.B.A. training.) Understanding the demand for mathematical science skills per se is critical for two reasons: (1) the demand for those skills, especially where it is increasing or moving in new directions, requires college and university education that in part relies on academic mathematical scientists and (2) the demand for those skills implies at least the possibility that the nation would benefit from a larger number of master's or Ph.D.-level mathematical scientists, especially if their training were designed with consideration of the emerging career options.

Dr. Simons noted that Renaissance Technologies has carried out all of its trading through quantitative models since 1998; it now works with several terabytes of data per day from markets around the world. The company has about 90 Ph.D.s, most in the research group but some in the programming group. Some of these people have backgrounds in mathematics, but others have degrees in astronomy, computer science, physics, or other disciplines that provide strong mathematical science skills. He listed a number of financial functions that rely on mathematical science skills: prediction, valuation, portfolio construction, volatility modeling, and so on. Even though the financial sector hires a large number of people with strong mathematical science expertise, he thinks the level of mathematical knowledge in the finance world is still lower than it should be. As an example, he said that many people do not know the distinction between beta (the difference between an instrument's performance and that of a relevant market) and volatility; they are related but different. He thinks finance will continue to be permeated by quantitative methods. Some of the skills that are necessary, in Dr. Simons's view, include statistics (though not normally at the level of new research) and optimization, and good programming skill is essential.

Dr. Simons is concerned about the pool of U.S.-born people with strong skills in the mathematical sciences. The majority of people hired by Renaissance are non-Americans. Most are from Europe, China, and India, though most have gone through a U.S. graduate program, and the fraction of U.S.-born people hired is declining. He thinks he probably could have found an adequate number of U.S.-born people if pressed, but it would have required a lot of work. He worries that high school teaching in the United States is simply not good enough, even though our economy is increasingly dependent on mathematical models and data analysis.

Dr. Dietrich described the kinds of mathematical science opportunities she sees and the kind of people IBM-Watson would like to hire. She said that much of IBM's business has become data-intensive, and numerical literacy is needed throughout the corporation. The mathematical sciences are increasingly central to economics, finance, business, and marketing, including areas such as risk assessment, game theory, and machine learning for marketing. But she noted that it is difficult to find enough people who have the ability to deal with large numerical data sets plus the ability to understand simple concepts such as range and variability. Many mathematical scientists at IBM must also operate as software developers, and they must be flexible enough to move from topic to topic.

She listed some qualifications that are especially valuable in her division, which employs over 300 people worldwide. It needs people with statistical expertise who are also are very computational. They should not rely on existing models and toolkits and should be comfortable working with messy data. The division needs people who are strong in discrete mathematics and able to extract understanding from big data sets. In her experience, most employees do not need to know calculus, and she would like to see more emphasis in the undergraduate curriculum on stochastic processes and large data. Programming ability is essential.

Mr. Bin Zafar presented impressions about the mathematical science skills that are important to the movie industry. (Similar skills are presumably important for the creation of computer games and computer-based training and simulation systems.) He showed the committee an example of a lengthy computer-generated sequence, from a major film release, that simulates the destruction of Los Angeles by a tsunami. Mathematical modeling was behind realistic images of wave motion and of building collapses, down to details such as the way windows would shatter and dust would rise and swirl. A great deal of effort is expended in creating tools for animation and computer-generated effects, both generic capabilities and particular instantiations.

Mr. Bin Zafar reported that of the several hundred people working in R&D at DreamWorks, about 13 percent have Ph.D.s and 34 percent have master's degrees. Just over half of the R&D staff have backgrounds in computer science, 19 percent in engineering, and 6 percent in a mathematical science. He mentioned that he does not receive many applications from mathematical scientists, and he speculated that perhaps they are not aware of the mathematical nature of work in the entertainment sector. He observed that creating robust and maintainable software is essential in his business—most software must be reliable enough to last perhaps 5 years—and that very few of their applicants develop that skill through education. Their schooling seems to assume that actual code creation is just an "implementation detail," but Mr. Bin Zafar observed that the implementation step often exposes very deep details that, if caught earlier, would have led the developer to take a different course.

Dr. Shum spoke first of his experience in helping Microsoft Research to establish a research laboratory in Beijing, beginning in 1999. He reported that there is plenty of raw talent in China, "every bit as good as MIT," so in setting up the research center in Beijing, a conscious effort was made to include some training opportunities that would enable the laboratory to develop that raw talent. By the time Dr. Shum left Beijing in 2006, Microsoft Research–Asia employed about 200 researchers, a few dozen postdoctoral researchers, and 250 junior workers.

Speaking more generally about Microsoft Research's needs, Shum mentioned three mathematical science areas that are of current importance to his search technology division of over 1,000 people:

- Auction theory, including mechanism design. The problem of mechanism design (see Chapter 2) is critical, and people with backgrounds in the mathematical theory are necessary. He has a few dozen people working on this topic.
- Graphs, including research that helps us manage enormous graphs such as Internet traffic patterns and research to understand social graphs, entity graphs, and click graphs (which show where users clicked on a hyperlink). His team at Microsoft Research includes many people with theoretical and mathematical backgrounds.

- Machine learning, which is a core foundation for advancing search technologies. His division includes perhaps 50 people working on aspects of machine learning.

More than half the people in Dr. Shum's division have backgrounds in computer science, and he has also hired engineers who have strong programming skills. Perhaps 5-10 percent of the people in his division received their final degrees in mathematics or statistics. Not too many of these people have Ph.D.s, though he has recently hired some Ph.D. statisticians.

This anecdotal information gathered by the committee is echoed in a more thorough examination by the McKinsey Global Institute.[1] That report estimated that U.S. businesses will need an additional 140,000-190,000 employees with "deep analytical talent" and a high level of quantitative skills by 2018. On p. 10, the report points out that "a significant constraint on realizing value from big data will be a shortage of talent, particularly of people with deep expertise in statistics and machine learning," to carry out analyses in support of corporate decision making. Preparing enough professionals to address this need constitutes both an opportunity and a challenge for the mathematical sciences. The careers examined by that report are those that deal with business analytics, especially as driven by large-scale data. Most of the people who fill those slots will need very strong mathematical science backgrounds, whether or not they actually receive a graduate degree in mathematics or statistics. And academic mathematical scientists must prepare to educate these additional people, regardless of what degrees they actually pursue. The McKinsey report goes on to say (p. 105) that "although we conducted this analysis in the United States, we believe that the shortage of deep analytical talent will be a global phenomenon. . . . Countries with a higher per capita production of deep analytical talent could potentially be attractive sources of these skills for other geographies either through immigration or through companies offshoring to meet their needs."

This McKinsey result supplements a well-known observation from Google's chief economist, Hal Varian, who was quoted in the *New York Times* as saying "the sexy job in the next 10 years will be statisticians . . . and I'm not kidding."[2] In addition to new careers spawned by the availability of large amounts of data and the information industry, many other fields—e.g., medicine—are also experiencing a growing need for professionals with sophisticated skills in the mathematical sciences.

---

[1] James Manyika, Michael Chui, Jacques Bughin, Brad Brown, Richard Dobbs, Charles Roxburgh, and Angela Hung Byers, 2011, *Big Data: The Next Frontier for Innovation, Competition, and Productivity*. McKinsey Global Institute, San Francisco.

[2] "For Today's Graduate, Just One Word: Statistics," *New York Times*, August 5, 2009.

Throughout its study, the committee heard many expressions of concern about the supply of home-grown talent in the mathematical sciences. That is, of course, a concern shared across all STEM disciplines. For a long time, the U.S. STEM workforce has been dependent on the flow of talented young people from other countries and on the fact that many of them are interested in building careers in the United States. Our country cannot depend on that situation continuing.

In recent years, there has been a great advance in our ability to quantify. But even top undergraduates too often have little or no experience and intuition about probability or concepts such as the central limit theorem, the law of large numbers, or indeterminacy. In order to prepare students for today's opportunities in the mathematical sciences, we need to push earlier on these skills. Our high schools focus on getting people prepared for calculus, and that influences even the elementary school curriculum. But we do little to teach statistics, probability, and uncertainty, instead acting as though students can just pick this up in the course of other learning. This is one of the biggest issues facing U.S. mathematical sciences; it is also a big problem in terms of national competitiveness.

The statistics profession might learn from the physics profession's attitude with regard to training. Physicists who have been trained as theoreticians may often then gain postdoctoral experience in experimental work (often in a different field). But statisticians are more rigid in their attitudes. For example, it is rare that statistics departments embrace a broad range of theoreticians, applied statisticians, and experimentalists. The latter category is important: When statisticians collect their own data, as some do, they are less likely to be relegated to supporting roles in scientific investigations, as can sometimes happen. Students educated in such an environment would have innate understanding of how to work in an interdisciplinary setting. However, statisticians may have difficulty obtaining funding to support data collection, and so the field cannot change in this direction unless funding programs evolve as well.

The mathematical sciences community plays a critical role in educating a broad range of students. Some will exhibit a special talent in mathematics from a young age and may remind successful researchers of their youthful selves. But there are many more whose interest in the mathematical sciences arises later and perhaps through nontraditional pathways, and these latter students constitute a valuable pool of potential majors and graduate students. A third cadre consists of students from other disciplines who need strong mathematical sciences education. All three pools of students need expert guidance and mentoring from successful mathematical scientists, and their needs are not identical. The mathematical sciences must successfully attract and serve all three of these cadres of students.

The challenge of producing an adequate number of people for STEM careers is of interest far beyond the mathematical sciences, of course. For example, a recent report[3] from the President's Council of Advisors on Science and Technology (PCAST) "provides a strategy for improving STEM education during the first two years of college that [it believes] is responsive to both the challenges and the opportunities that this crucial stage in the STEM education pathway presents," according to the cover letter to the President that accompanies the report.[4] That cover letter goes on to recount the reason why STEM fields receive this attention:

> Economic forecasts point to a need for producing, over the next decade, approximately 1 million more college graduates in STEM fields than expected under current assumptions. Fewer than 40% of students who enter college intending to major in a STEM field complete a STEM degree. Merely increasing the retention of STEM majors from 40% to 50% would generate three-quarters of the targeted 1 million additional STEM degrees over the next decade.[5]

That PCAST report goes on to recommend, among other steps, a "multi-campus 5-year initiative aimed at developing new approaches to remove or reduce the mathematics bottleneck that is currently keeping many students from pursuing STEM majors." This proposed initiative might involve approximately 200 "experiments" exploring a variety of approaches, including the following:

(1) Summer and other bridge programs for high school students entering college;
(2) remedial courses for students in college, including approaches that rely on computer technology;
(3) college mathematics teaching and curricula developed and taught by faculty from mathematics-intensive disciplines other than mathematics, including physics, engineering, and computer science; and
(4) a new pipeline for producing K-12 mathematics teachers from undergraduate and graduate programs in mathematics-intensive fields other than mathematics.

It is critical that the mathematical sciences community actively engage with STEM discussions going on outside the mathematical sciences com-

---

[3] PCAST, 2012, *Engage to Excel: Producing One Million Additional College Graduates with Degrees in Science, Technology, Engineering, and Mathematics*. The White House, Washington, D.C.

[4] Available at http://www.whitehouse.gov/sites/default/files/microsites/ostp/pcast-engage-to-excel-final_2-25-12.pdf.

[5] Ibid.

munity and not be marginalized in efforts to improve STEM education, especially since those plans would greatly affect the responsibilities of mathematics and statistics faculty members. This committee knows of no evidence that teaching lower-division college mathematics and statistics or providing a mathematical background for K-12 mathematics teachers can be done better by faculty from other subjects but it is clear that the mathematics-intensive disciplines are full of creative people who constitute a valuable resource for innovative teaching ideas. The need to create a truly compelling menu of creatively taught lower-division courses in the mathematical sciences tailored to the needs of twenty-first century students is pressing, and partnerships with mathematics-intensive disciplines in designing such courses are eminently worth pursuing.

The traditional lecture-homework-exam format that often prevails in lower-division mathematics courses would benefit from a reexamination. One aspect of the changes PCAST would like to see is explained in its report:

> Better teaching methods are needed by university faculty to make courses more inspiring, provide more help to students facing mathematical challenges, and to create an atmosphere of a community of STEM learners. Traditional teaching methods have trained many STEM professionals, including most of the current STEM workforce. But a large and growing body of research indicates that STEM education can be substantially improved through a diversification of teaching methods. These data show that evidence-based teaching methods are more effective in reaching all students—especially the 'underrepresented majority'—the women and members of minority groups who now constitute approximately 70% of college students while being underrepresented among students who receive undergraduate STEM degrees (approximately 45%).

In an appendix to the report, some of the methods the PCAST working group would like to see explored are (1) active learning techniques; (2) motivating learning by explaining how mathematics is used and making courses more relevant for students' fields of specialization; (3) creating a community of high expectations among students; and (4) expanding opportunities for undergraduate research.

The PCAST report should be viewed as a wake-up call for the mathematical sciences community. While there have been numerous promising experiments within the community for addressing the issues it raises—especially noteworthy has been the tremendous expansion in opportunities for undergraduate research in the mathematical sciences—at this point a community-wide effort is called for. The professional societies should work cooperatively to spark this. Change is unquestionably coming to lower-division undergraduate mathematics, and it is incumbent upon the mathematical sciences community to ensure that it is at the center of these changes and not at the periphery.

THE TYPICAL EDUCATIONAL PATH IN THE MATHEMATICAL SCIENCES NEEDS ADJUSTMENTS

Chapter 3 showed exciting emerging opportunities for anyone with expertise in the mathematical sciences. The precise thinking and conceptual abilities that are hallmarks of mathematical science training continue to be an excellent preparation for many career paths. However, it is apparent that an ability to work with data and computers is a common need. An understanding of statistics, probability, randomness, algorithms, and discrete mathematics are probably of greater importance than calculus for many students who will follow such careers, and indeed students with this training will be much more employable in those areas. The educational offerings of typical departments in the mathematical sciences have not kept pace with the changes in how the mathematical sciences are used. A redesigned offering of courses and majors is needed. Although there are promising experiments, a community-wide effort is needed in the mathematical sciences to make its undergraduate courses more compelling to students and better aligned with needs of user departments.

The 2012 report of the Society for Industrial and Applied Mathematics (SIAM) on mathematical sciences in industry[6] adds support for this, with regard to those students who would like to work in industry. The SIAM report makes the following statement about useful background for such people:

> Useful mathematical skills include a broad training in the core of mathematics, statistics, mathematical modeling, and numerical simulation, as well as depth in an appropriate specialty. Computational skills include, at a minimum, experience in programming in one or more languages. Specific requirements, such as C++, a fourth-generation language such as MATLAB, or a scripting language such as Python, vary a great deal from company to company and industry to industry. Familiarity with high-performance computing (e.g., parallel computing, large-scale data mining, and visualization) is becoming more and more of an asset, and in some jobs is a requirement. . . . In general, the student's level of knowledge [of an application domain] has to be sufficient to understand the language of that domain and bridge the gap between theory and practical implementation.[7]

---

[6] SIAM, 2012, *Mathematics in Industry*. Society for Industrial and Applied Mathematics, Philadelphia, Pa. Available at http://www.siam.org/reports/mii/2012/index.php.

[7] Ibid., p. 2 of Summary.

TABLE 5-1 Enrollment (in 1000s) in Undergraduate Courses Taught in the Mathematics or Statistics Departments of Four-Year Colleges and Universities, and in Mathematics Programs of Two-Year Colleges, for Fall 1990, 1995, 2000, 2005, and 2010

Discipline	Fall 1990	Fall 1995	Fall 2000	Fall 2005	Fall 2010
Mathematics	1,621	1,471	1,614	1,607	1,971
Statistics	169	208	245	260	371

SOURCES: CBMS, 2007, 2012.

It may be that students are already "voting with their feet." According to data from the Conference Board of the Mathematical Sciences (CBMS), the number of enrollments in mathematics courses in 1990-2010 remained generally flat, while the number of enrollments in statistics courses increased by 120 percent.[8,9] The raw numbers are shown in Table 5-1.

The four industry leaders who spoke with the committee, and whose observations were recounted earlier in this chapter, raised the need for more people who focus on real problems—rather than on models that omit too much of the messiness of reality—and who are able to work with computers, statistics, and data so as to test and validate their modeling. Theory alone is not the best preparation for, say, careers at IBM, Renaissance Technologies, or DreamWorks Animation. But an optimistic lesson to draw from these discussions is that industry has an increasing need for students with mathematical science skills, whether or not the skills are explicitly labeled that way.

The role of the mathematical sciences in science, engineering, medicine, finance, social science, and society at large has changed enormously, at a pace that challenges the university mathematical sciences curriculum. This change necessitates new courses, new majors, new programs, and new educational partnerships with those in other disciplines, both inside and outside universities. New educational pathways for training in the mathematical sciences need to be created—for students in mathematical sciences departments, for those pursuing degrees in science, medicine, engineering, business, and social science, and for those already in the workforce needing additional quantitative skills. New credentials may be needed, such as professional master's degrees for those about to enter the workforce or

---

[8] CBMS, 2007, *Statistical Abstract of Undergraduate Programs in the Mathematical Sciences in the United States; Fall 2005 CBMS Survey*, Table S.1. Available at http://www.ams.org/profession/data/cbms-survey/full-report.pdf.

[9] CBMS, 2012, *Draft of the Statistical Abstract of Undergraduate Programs in the Mathematical Sciences in the United States; Fall 2010 CBMS Survey*. Table S.1. Available at http://www.ams.org/profession/ data/cbms-survey/cbms2010-work.

already in it. The trend toward periodic acquisition of new job skills by those already in the workforce provides an opportunity for the mathematical sciences to serve new needs.

Most mathematics departments still tend to use calculus as the gateway to higher-level coursework, and that is not appropriate for many students. Although there is a very long history of discussions about this issue, the need for a serious reexamination is real, driven by changes in how the mathematical sciences are being used. For example, someone who wants to study bioinformatics ought to have a pathway whereby he or she can learn probability and statistics; learn enough calculus to find maxima and minima and understand ordinary differential equations, get a solid dose of discrete mathematics; learn linear algebra; and get an introduction to algorithms. Space could be made in their curriculum by deemphasizing such topics as line integrals and Stokes's theorem, epsilons and deltas, abstract vector spaces, and so on. Different pathways are needed for students who may go on to work in bioinformatics, ecology, medicine, computing, and so on. It is not enough to rearrange existing courses to create alternative curricula. As one step in this direction, colleges and universities might encourage AP statistics courses as much as they do AP calculus. Such a move would also help those in secondary education who believe that teaching of probability, statistics, and uncertainty should be more common.

The dramatic increase over the past 20 years in the number of NSF Research Experience for Undergraduate (REU) programs has, in the experience of members of the study committee, been a noticeable force for attracting talented undergraduates to major in a mathematical science while also providing a stronger foundation for graduate study.[10] Another striking trend over the past decade or two is the increase in double majors. This increase means that undergraduates who might otherwise pursue a nonmathematical sciences major gain exposure to a broader array of mathematics and statistics courses and, in essence, keep more career options open. Double or flexible majors have also enabled some departments in mathematics and statistics to increase the number of undergraduates in their programs and keep them strongly involved at least through their bachelor's degrees.

Many graduate students will end up not with traditional academic jobs but with jobs where they are expected to deal with problems much less well formulated than those in the academic setting. They must bring their math-

---

[10] PCAST, 2012, *Engage to Excel: Producing One Million Additional College Graduates with Degrees in Science, Technology, Engineering, and Mathematics.* The White House, Washington, D.C. Appendix G of this report recounts some anecdotal evidence of the value of undergraduate research experiences for building student commitment to STEM fields and retaining it.

ematical sciences talent and sophistication to bear on ill-posed problems so as to make a contribution to the solution of these problems. This requires different skills from the ones that they trained for during their graduate student days, and it suggests that the training of graduate students in the mathematical sciences needs to be rethought given the changing landscape in which students may now work. At the least, mathematics and statistics departments should take steps to ensure that their graduate students have a broad and up-to-date understanding of the expansive reach of the mathematical sciences.

> Recommendation 5-1: Mathematics and statistics departments, in concert with their university administrations, should engage in a deep rethinking of the different types of students they are attracting and wish to attract and must identify the top priorities for educating these students. This should be done for bachelor's, master's, and Ph.D.-level curricula. In some cases, this rethinking should be carried out in consultation with faculty from other relevant disciplines.

> Recommendation 5-2: In order to motivate students and show the full value of the material, it is essential that educators explain to their K-12 and undergraduate students how the mathematical science topics they are teaching are used and the careers that make use of them. Modest steps in this direction could lead to greater success in attracting and retaining students in mathematical sciences courses. Graduate students should be taught about the uses of the mathematical sciences so that they can pass this information along to students when they become faculty members. Mathematical science professional societies and funding agencies should play a role in developing programs to give faculty members the tools to teach in this way.

The mathematical science community collectively does not do a good job in its interface with the general public or even with the broader scientific community, and improving this would contribute to the goal of enlarging the STEM pipeline. Internet tools such as blogs and video lectures offer new pathways for this outreach, which may be appealing to both practicing and retired mathematical scientists. There is a special need to improve the general level of understanding about uncertainty, which relies on an understanding of probability and statistics.

> Recommendation 5-3: More professional mathematical scientists should become involved in explaining the nature of the mathematical sciences enterprise and its extraordinary impact on society. Academic departments should find ways to reward such work. Professional soci-

eties should expand existing efforts and work with funding entities to create an organizational structure whose goal is to publicize advances in the mathematical sciences.

Finally, the committee notes that the boom-and-bust cycles of the academic job market, especially for new Ph.D.s, result in a substantial loss of talent because they both discourage entry to research in the mathematical sciences and increase the likelihood of exit from it. The impact on core mathematical sciences, where the academic job market is central, is especially severe. Important workforce programs, such as NSF's former VIGRE program, are often dwarfed by these macroeconomic trends. Stabilizing these swings by expanding the availability of postdoctoral fellowships during downturns in the job market should be an important component of the nation's overall strategy to strengthen the mathematical sciences workforce and ensure continuity over long time horizons. NSF/DMS did just that during the recession of 2008-2009, and it would be ideal if a mechanism were in place to respond similarly during the next downturn in hiring.

It is because of the importance and centrality of the mathematical sciences, as detailed elsewhere in this report, that these educational issues are as important as they are. As a community, mathematical scientists have been handed an extraordinary opportunity to play a central role in educating researchers and professionals in many of the most exciting career and research areas of the twenty-first century. Taking advantage of this opportunity requires a certain amount of cultural flexibility and the development of educational partnerships with those in other disciplines. The benefits to the country and to the mathematical sciences profession would be enormous.

Appendix C provides additional basic data about employment and Ph.D. production in the mathematical sciences.

ATTRACTING MORE WOMEN AND UNDERREPRESENTED MINORITIES TO THE MATHEMATICAL SCIENCES

Concerns About the Current Demographics

The underrepresentation of women and ethnic minorities in mathematics has been a persistent problem for the field. Fifty years ago, the mathematical sciences community consisted almost exclusively of white males, and that segment of the population remains the dominant one from which the community attracts new members. This implies that talent in other sectors of the population is being underutilized, and as white males become a smaller fraction of the population, it is even more essential that the mathematical sciences attract and retain students from across the totality of the population. While there has been significant progress in the last

10-20 years, the fraction of women and minorities in the mathematical sciences drops with each step along the pipeline and up the career ladder. This very leaky pipeline, which was identified as a problem in the 1990 "David II" report[11] and earlier, is now the key problem in achieving further diversity and undermines the ability of the mathematical sciences to make full use of its potential talent pool. This section briefly examines the current state of minority and female representation at various levels (K-12, undergraduate, master's, Ph.D., and the professoriate) along with recent trends, and it profiles some efforts that are under way to encourage greater representation.

In elementary school, girls perform much like boys on mathematics standardized testing. Standardized testing scores indicate that young girls (age 9) are performing at the same level—if not a higher one—than boys of the same age. However, a score gap between girls and boys appears in middle school (age 13) and grows in high school (age 17). A contrasting and rather revealing study of this issue appeared in a 2008 article that studied the effects of culture on the participation of girls at the International Mathematical Olympiad (IMO) teams among children from different countries.[12] The authors found that, based on IMO participation, some East European and Asian countries produce girls with profound ability in mathematical problem solving; most other countries, including the United States, do not. Further, they found that girls on the U.S. team often are recent immigrants from countries that typically produce such talented girls. While they do not identify the environmental factors that make these countries more supportive of girls, the study shows the strong effect of the environment on bringing out mathematical talent in girls, and it suggests that the United States can do a lot more to avoid wasting this talent, as discussed below.

In the United States, approximately 40 percent of the bachelor's degrees awarded in the mathematical sciences are awarded to women. Because more women than men now attend college, there is definitely room for improvement. Although this rate of female participation is enviable compared to rates found in many other technical fields, there is still a lost opportunity, because more women than men drop out of the mathematical sciences pipeline after high school. Then in college, while mathematics initially attracts as many women as men, women seem to move away from the field at a higher rate before graduation. In particular, graduate training in mathematics clearly does not attract as many women as men. Of the total

---

[11] NRC, 1990, *Renewing U.S. Mathematics: A Plan for the 1990s.* National Academy Press, Washington, D.C.

[12] Titu Andreescu, Joseph A. Gallian, Jonathan M. Kane, and Janet E. Mertz, 2008, Cross-cultural analysis of students with exceptional talent in mathematical problem solving. *Notices of the AMS* 55(10):1248-1260.

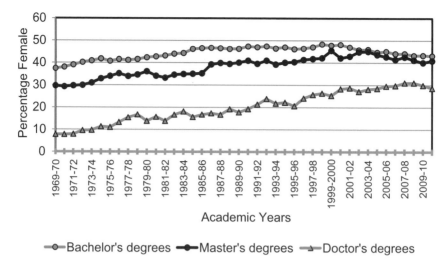

FIGURE 5-1 Degrees in mathematics and statistics conferred by degree-granting institutions, by level of degree and sex of student, 1969-1970 through 2010-2011. SOURCE: National Center for Education Statistics, 2012, *Digest for Education Statistics*. Table 327. Available at http://nces.ed.gov/programs/digest/d12/tables/dt12_327.asp.

doctorates granted in 2009-2010 (1,632), 31 percent of recipients were female. Figure 5-1 shows the trends in the percentage of females receiving bachelor's, master's, and Ph.D. degrees in mathematics and statistics from 1969 to 2009.[13]

It is interesting to note that, while the fraction of mathematics Ph.D.s awarded to women is about 30 percent, in recent years women have been awarded more than half the statistics Ph.D.s in the United States. The fact that statistics attracts a higher percentage of women than mathematics is worth understanding better.

Within the professoriate, the data get more complicated. As shown in Table 5-2, the percentage of women among full-time faculty at 4-year institutions rose to 26 percent in mathematics departments and 22 percent in statistics departments in 2005, when the most recent CBMS data were collected. However, this "full-time" status can be used to describe different positions, and women tended to be disproportionately represented in positions that were not tenured or tenure-track in 2005.[14]

---

[13] From *Digest of Education Statistics*, 2010, Table 323. Available at http://nces.ed.gov/programs/digest/d10/tables/dt10_323.asp.

[14] More recent data are in being assembled by CBMS but have not yet been published.

TABLE 5-2 Gender of Full-Time Faculty Members in Departments of Mathematics and Statistics at 4-Year Colleges and Universities and in Departments of Statistics at Universities

Department	Number of	1975	1980	1985	1990	1995	2000	2005	2010
Mathematics	Full-time faculty	16,863	16,022	17,849	19,411	18,248	19,779	21,885	22,294
	Women	1,686 (10%)	2,243 (14%)	2,677 (15%)	3,843 (20%)	3,880 (21%)	4,346 (22%)	5,641 (26%)	6,417 (29%)
Statistics	Full-time faculty	NA	NA	740	735	988	1,022	946	1,265
	Women	NA	NA	74 (10%)	105 (14%)	107 (11%)	179 (18%)	211 (22%)	327 (26%)

NOTE: 1975, 1980, 1985, 1990, and 1995 data from CBMS, 1997, *Statistical Abstract of Undergraduate Programs in the Mathematical Sciences in the United States, Fall 1995 CBMS Survey,* Table SF8, available at http://www.ams.org/profession/data/cbms-survey/cbms1995; 2000 and 2005 data from CBMS, 2007, Table F.1; 2010 data from CBMS, 2012, Table F.1.

Another consideration is what type of 4-year institution employed these women. Data in Appendix C show that women are much less represented in Ph.D.-granting universities than in other types of 4-year institutions. However, there are some indications that in the last 5 years or so all kinds of universities made good progress in increasing the percentage of women who were hired.[15] Still, the mathematical sciences are not retaining as many women in the pipeline as would be desired, and, in particular, not enough women are being hired into academic careers.[16]

Several racial and ethnic groups (most notably black, Hispanic, and Native American/Alaska Native) are even more seriously underrepresented in the mathematical sciences. Standardized mathematics tests in K-12 show a performance gap between whites and both blacks and Hispanics, with a notable score gap existing between whites and blacks through elementary, middle, and high school. Data show that there is also a notable and persistent score gap between whites and Hispanics. However, the score gap has been decreasing at the elementary school level for Hispanics over the past decade. It is important to note that these populations do have an interest in STEM subjects. A recent report from the National Academies points out as follows:

> Recent data from the Higher Education Research Institute (HERI) at UCLA show that underrepresented minorities aspire to major in STEM in college at the same rates as their white and Asian American peers, and have done so since the late 1980s.[17]

At the college and university level, only 5.7 percent and 6.4 percent of bachelor's degrees in mathematics and statistics are awarded to blacks and Hispanics, respectively. This underrepresentation continues in graduate school, where blacks make up only 2.9 percent of master's degree recipients and 2.0 percent of Ph.D. recipients. Hispanics make up 3.1 percent of master's degree recipients and 2.5 percent of the Ph.D. recipients. American

---

[15] A recent report from MIT, *Report on the Status of Women Faculty in the Schools of Science and Engineering at MIT*, 2011, recounts that university's progress on this issue over the last 10 years and the important remaining issues it faces. Available at http://web.mit.edu/newsoffice/images/documents/women-report-2011.pdf.

[16] A recent article—Corinne A. Moss-Racusin, John F. Dovidio, Victoria L. Brescoll, Mark J. Graham, and Jo Handelsman, 2012, Science faculty's subtle gender biases favor male students, *Proceedings of the National Academy of Sciences* 109 (September 17)—suggests that some subtle and pervasive cultural biases may be at play and need to be countered.

[17] Institute of Medicine, National Academy of Sciences, and National Academy of Engineering, 2011, *Expanding Underrepresented Minority Participation: America's Science and Technology Talent at the Crossroads*. The National Academies Press, Washington, D.C., p. 4, based on UCLA Higher Education Research Institute, 2010, *Degrees of Success: Bachelor's Degree Completion Rates Among Initial STEM Majors*. HERI Report Brief, January 2010.

TABLE 5-3 Proportion of Full-Time Faculty Belonging to Various Ethnic Groups, by Gender and Type of Department, in Fall 2005

Department	Asian	Black, Not Hispanic	Mexican American/ Puerto Rican/ Other Hispanic	White, Not Hispanic	Other/ Unknown[a]
Mathematics					
Ph.D.-awarding					
Men	13	1	2	59	3
Women	4	0	1	16	1
Master's-awarding					
Men	12	4	2	47	2
Women	5	2	1	26	1
Bachelor's-awarding					
Men	4	2	2	57	2
Women	2	1	1	28	1
Statistics					
Men	20	1	1	49	3
Women	8	0	1	15	2

NOTE: Zero means less than 0.5 percent. Except for round-off, the percentages within each departmental type sum to 100.0. SOURCE: CBMS, 2012, Table F.5.

[a] The column "Other/Unknown" includes the federal categories Native American/Alaskan Native and Native Hawaiian/Other Pacific Islander.

Indians/Alaska natives make up, respectively, 0.5, 0.2, and 0.2 percent of bachelor's, master's, and Ph.D. recipients.[18]

Table 5-3 gives a high-level view of the participation of traditionally underrepresented minorities among full-time faculty in all mathematics departments and in Ph.D.-awarding statistics departments.

According to AMS data,[19] the number of U.S.-citizen Ph.D.s from all under-represented minority groups was 12 in 1992, 27 in 2000, and 45 in 2011. (The total number of U.S. citizens receiving Ph.D.s from mathematics or statistics departments in 2011 was 802.) It is perhaps gratifying to see an approximate doubling each decade, but the total numbers are still strikingly small. Moreover, only 11 of the Ph.D.s in 2011 were awarded from Group I departments. For comparison, in 2011 Group I departments awarded a total of 272 Ph.D.s to U.S. citizens and 560 Ph.D.s overall.

In order for the nation to have a workforce that can exploit the kinds of opportunities described in Chapters 3 and 4, the mathematical sciences

---

[18] See Figures C-4, C-5, and C-6 in Appendix C.

[19] Available at http://www.ams.org/profession/data/annual-survey/2011Survey-SREC.pdf.

enterprise must improve its ability to attract and retain a greater fraction of talented young people. As indicated in the introduction to this chapter, this is a high-priority national issue.

What Can Be Done?

There have been some notable successes in attracting and retaining more under-represented minorities in the mathematical sciences. For example, William Vélez of the University of Arizona at Tucson has successfully increased minority enrollment. He offered the following advice for recruiting all types of students:

- Provide timely information to students. Help them to understand the system and future opportunities. Even good students need attention and advice.
- Examine ways to ease the transition from high school to college or university.
- Encourage students who are interested in science and engineering to have a second major in mathematics.
- Pay more individual attention to talented students by having faculty reach out to them directly.
- Communicate the necessity of studying mathematics.[20]

While these suggestions are not unique, the practices are often not implemented. They can be broadly applied to all students, regardless of race or gender, to increase the population of undergraduate majors in the mathematical sciences.

Despite the small numbers of underrepresented minorities entering the mathematical sciences, there are a number of programs across the country that are quite successful at achieving greater participation. They have established practices that work and which could be replicated elsewhere. A recent report from the National Academies[21] presents a thorough examination of approaches for tapping this talent.

The NSF-supported mathematical science institutes have also been active in efforts to reach out to underrepresented groups. For example, the Institute for Mathematics and its Applications (IMA) and the Institute for Pure and Applied Mathematics (IPAM) offer workshops in professional devel-

---

[20] William Yslas Vélez, 2006, "Increasing the number of mathematics majors," *FOCUS Newsletter*, Mathematical Association of America, March.

[21] Institute of Medicine, National Academy of Sciences, and National Academy of Engineering, 2011, *Expanding Underrepresented Minority Participation: America's Science and Technology Talent at the Crossroads*. The National Academies Press, Washington, D.C.

opment aimed at mathematical scientists from under-represented groups. At the K-12 level, IPAM, IMA, and other institutes have offered week-long programs for middle and high school girls. In rotation, the institutes offer the Blackwell-Tapia conferences, which aim to increase the exposure of underrepresented groups to mathematics. Some efforts of the Mathematical Sciences Research Institute (MSRI) aim at increasing the participation of women and minorities:

- Connections for Women workshops, 2-day workshops that aim to showcase women's talent in the field and that sometimes offer an intensive minicourse on fundamental ideas and techniques;
- MSRI-UP, a program for undergraduates aimed at increasing the participation of underrepresented groups in mathematics graduate programs; and
- The Network Tree, a project to compile names and contact information for mathematicians from underrepresented groups.

Colette Patt from the Science Diversity Office of the University of California, Berkeley, and Deborah Nolan and Bin Yu from the Statistics Department at that university shared with the committee the following lists of issues (adapted by the committee) they compiled that academic departments should consider when determining how to improve their recruitment and retention of women and other underrepresented groups.

Issues That Affect Recruitment and Retention at the Undergraduate Level

- Affordability of undergraduate education and awareness of assistance programs, such as Research Experiences for Undergraduates and support for travel to conferences;
- Awareness of and motivation to enter the mathematical sciences, such as information about career options made possible by mathematical science coursework or majors and comparison of those options to some more common career paths;
- Adequacy of mentoring, including encouragement, coaching, and strategic advising;
- Access to, and encouragement to participate in, a variety of research opportunities;
- The possibility of boosting confidence by departmental approaches to structuring the curriculum and course pedagogies, such as confidence, study habits, sense of community, and so on;
- Academic requirements, structure of courses and majors, academic support, choice of gateway courses, teaching effectiveness, and classroom practices;
- Campus climate and department culture.

Issues That Affect Recruitment and Retention at the Graduate Level

- Availability of role models;
- Need for a sense of belonging and community to avoid possible isolation;
- Possible harassment, peer interactions, and climate issues;
- Availability and skill of mentoring;
- Opportunities for professional development and socialization;
- Psychological factors that possibly can be boosted by departments' approaches to structuring the graduate curriculum, courses, and tests to influence factors such as confidence, self-concept, science identity, and the threats of being stereotyped.
- Monitoring and possible intervention to assist at the critical transition from the graduate to postdoctoral positions;
- Assistance in goal-setting and evaluation.

Issues That Affect Recruitment and Retention of Underrepresented Faculty

- Understanding and countering the drop-off of women and minorities at the critical transition from postdoctoral years to faculty careers;
- Understanding and countering the difficulties of achieving a life-work balance, which tends to affect women more than men;
- Identifying perceptions that are gender differentiated and can affect seemingly objective measures—for example, gender bias in letters of recommendation, teaching evaluations, perceptions of leaders;
- Opportunities for leadership;
- Differential recognition, awards, and the accumulation of cultural capital in the field.

Many of these issues have been the subject of published studies that document their impact on the recruitment and retention of women and other underrepresented groups, and most should be familiar to anyone who has spent time in academic departments.

Statistics departments have been quite successful in recent years in attracting and retaining women, and it would be very helpful to understand better how the broader mathematical sciences community can learn from this success. A similar observation has been made with regard to attracting women to application-oriented computer science (CS) programs.[22]

Overall, there has been progress in attracting women and minorities to the mathematical sciences. Unfortunately, the accumulation of small disadvantages women and minorities face throughout their career can add up to a significant disadvantage and can cause the leaking of the pipeline that

---

[22] See Christine Alvarado and Zachary Dodds, 2010, Women in CS: An evaluation of three promising practices. *Proceedings of SIGCSE 2010* March 10-13. Association for Computing Machinery, Milwaukee, Wisc.

is documented above. Beyond this, one or more egregious incidents can tip the balance for an individual. This is an important issue for the mathematical sciences to address.

> **Recommendation 5-4: Every academic department in the mathematical sciences should explicitly incorporate recruitment and retention of women and underrepresented groups into the responsibilities of the faculty members in charge of the undergraduate program, graduate program, and faculty hiring and promotion. Resources need to be provided to enable departments to monitor and adapt successful recruiting and mentoring programs that have been pioneered at many schools and to find and correct any disincentives that may exist in the department.**

Appendix E lists some of the organizations and programs that are committed to improving participation by women and minorities in the mathematical sciences at all levels of education.

THE CRITICAL ROLE OF K-12 MATHEMATICS AND STATISTICS EDUCATION

The extent to which size of the pipeline of students preparing for mathematical science-based careers can be enlarged is fundamentally limited by the quality of K-12 mathematics and statistics education. The nation's well-being is dependent on a strong flow of talented students into careers in STEM fields, but college students cannot even contemplate those careers unless they have strong K-12 preparation in the mathematical sciences. Absent such preparation, most are unlikely to be interested. Those statements are even more apt with respect to young people who could become mathematical scientists per se. The K-12 pipeline is an Achilles heel for U.S. innovation. Fortunately, a lot of innovation is taking place in K-12 mathematics and statistics education, and the mathematical sciences community has a role to play in strengthening and implementing the best of these efforts. This section gives a brief overview of the issues and pointers to the relevant literature. It is beyond the mandate of the current study to recommend actions in response to this general national challenge.

There are a large number of K-12 schools, both public and private, that perform at a high level year after year across the United States. Annual rankings of the best U.S. high schools document the top few on the basis of student performance parameters and other criteria.[23] Most states employ

---

[23]For example, *US News and World Report,* America's Best High Schools, November 29, 2007; *Newsweek,* Best High Schools in the U.S., June 19, 2011; *Bloomberg Business Week,* America's Best High Schools 2009, January 15, 2009.

information systems that keep detailed public school records for students, teachers, and school administrators on the basis of parameters established for mandatory statewide use. Public schools are subject to state-enforced sanctions when a school fails to meet the mandated performance criteria. But overall, particularly in the sciences and mathematics, U.S. K-12 students continue to perform substantially below average in international comparisons.

Education Secretary Arne Duncan's report on December 7, 2010, presented on the occasion of the release of the 2009 results of the Program for International Student Assessment (PISA) of the Organisation for Economic Co-operation and Development (OECD), did not contain encouraging news about the performance of U.S. 15-year-olds in mathematics.[24] U.S. students ranked 25th among the 34 participating OECD nations, the same level of performance as 6 years earlier in 2003. The results were not encouraging in reading literacy either, with U.S. students placing 14th, effectively no change since 2000. The only improvement noted was a 17th place ranking in science, marginally better than the 2006 ranking. Secretary Duncan added that the OECD analysis suggests the 15-year-olds in South Korea and Finland are, on average, 1 or 2 years ahead of their American peers in math and science.

The picture is not improving. In September 2011, the College Board reported that the SAT scores for the U.S. high school graduating classes of 2011 fell in all three subject areas tested: reading, writing, and mathematics. The writing scores were the lowest ever recorded.[25] A report from Harvard's Program on Education Policy and Governance in August of 2011 revealed that U.S. high school students in the Class of 2011 ranked 32nd in mathematics among OECD nations that participated in PISA for students at age 15. The report noted that 22 countries significantly outperform the United States in the share of students who reach the "proficient" level in math (a considerably lower standard of performance than "advanced").[26]

In September 2007 McKinsey & Co. produced what it called a first-of-its-kind approach that links quantitative results with qualitative insights on what high-performing and rapidly improving school systems have in common.[27] McKinsey studied 25 of the world's school systems, including 10 of the top performers. They examined what high-performing school systems have in common and what tools they use to improve student outcomes. They concluded that, overall, the following matter most:

---

[24] Available at http://www.ED.gov, December 7, 2010.

[25] *Wall Street Journal*, "SAT Reading, Writing Scores Hit New Low," September 15, 2011.

[26] Paul E. Peterson, Ludgar Woessmann, Eric A. Hanushek, and Carlos X. Lastra-Anadon, 2011, *Globally Challenged: Are U.S. Students Ready to Compete*. Harvard Kennedy School of Government, August.

[27] McKinsey & Co., 2007, *How the World's Best Performing School Systems Came Out on Top*.

- Getting the right people to become teachers (the quality of an education system cannot exceed the quality of its teachers);
- Developing them into effective instructors (the only way to improve outcomes is to improve instruction); and
- Ensuring that the system is able to provide the best possible instruction for every child (high performance requires every child to succeed).

The McKinsey report concludes: "The available evidence suggests that the main driver of the variation in student learning at school is the quality of the teachers." Three illustrations are provided to support this conclusion:

- Ten years ago, seminal research based on data from the Tennessee Comprehensive Assessment Program tests showed that if two average 8-year-old students were given different teachers—one of them a high performer, the other a low performer—the students' performance diverged by more than 50 percentile points within 3 years.[28]
- A study from Dallas showed that the performance gap between students assigned three effective teachers in a row and those assigned three ineffective teachers in a row was 49 percentile points.[29]
- In Boston, students placed with top-performing math teachers made substantial gains, while students placed with the worst teachers regressed—their math actually got worse.

The McKinsey report further concluded as follows:

Studies that take into account all of the available evidence on teacher effectiveness suggest that students placed with high-performing teachers will progress three times as fast as those placed with low-performing teachers.

The second McKinsey report (2010) addresses the teacher talent gap by examining the details of teacher preparation and performance in three top-performing countries: Singapore, Finland, and South Korea.[30] These

---

[28] W. Sanders and J. Rivers, 1996, *Cumulative and Residual Effects of Teachers on Future Student Academic Achievement.* University of Tennessee, Value-Added Research and Assessment Center, Knoxville, Tenn.

[29] Heather R. Jordan, Robert L. Mendro, and Dash Weerasinghe, 1997, "Teacher Effects on Longitudinal Student Achievement: A Report on Research in Progress," Presented at the CREATE Annual Meeting Indianapolis, Ind. Available at http://dallasisd.schoolwires.net/cms/lib/TX01001475/Centricity/Shared/evalacct/research/articles/Jordan-Teacher-Effects-on-Longitudinal-Student-Achievement-1997.pdf.

[30] Byron Auguste, Paul Kihn, and Matt Miller, 2010, "Closing the talent gap: Attracting and retaining top-third graduates to careers in teaching." McKinsey & Company, September.

three countries recruit 100 percent of their teacher corps from the top third of their college graduate academic cohort, then screen for other important qualities as well. By contrast, in the United States only 23 percent of new K-12 teachers come from the top third, and in high poverty schools, the fraction is only 14 percent. The report concludes that Finland, Singapore, and South Korea "use a rigorous selection process and teacher training more akin to medical school and residency than a typical American school of education." It goes on to examine what an American version of a "top third" strategy might entail and concludes that "if the U.S. is to close its achievement gap with the world's best education systems—and ease its own socio-economic disparities—a top-third strategy for the teaching profession must be a part of the debate." Undoubtedly, a part of closing this gap must address the situation that most teachers of mathematics and science in U.S. public middle and high schools do not have degrees or other certification in mathematics or science.[31]

ENRICHMENT FOR PRECOLLEGE STUDENTS WITH CLEAR TALENT IN MATHEMATICS AND STATISTICS

While, as noted above, the current study does not have a mandate to examine the broad question of K-12 mathematics education, the mathematical sciences community does have a clear interest in those precollege students with special talent for and interest in mathematics and statistics. Such students may very well go on to become future leaders of the research community, and in many cases they are ready to learn from active researchers while still in high school, or even earlier.

A 2010 paper[32] reported on two studies into the relationship between precollegiate advanced/enriched educational experiences and adult accomplishments in STEM fields. In the first of these studies, 1,467 13-year-olds were identified as mathematically talented on the basis of scores of at least 500 on the mathematics section of the Scholastic Assessment Test, which puts them in the top 0.5 percentile. Their developmental trajectories were studied over 25 years, with particular attention being paid to accomplishments in STEM fields, such as scholarly publications, Ph.D. attainment, tenure, patents, and types of occupation(s) over the period. The second study profiled, retrospectively, the adolescent advanced/enriched educa-

---

[31] NRC, 2010, *Rising Above the Gathering Storm, Revisited: Rapidly Approaching Category 5*. The National Academies Press, Washington, D.C.

[32] Jonathan Wai, David Lubenski, Camilla Benbow, and James Steiger, 2010, Accomplishment in science, technology, engineering, and mathematics (STEM) and its relation to STEM educational dose: A 25-year longitudinal study. *Journal of Educational Psychology* 102 (4).

tional experiences of 714 top STEM graduate students and related their experiences to their STEM accomplishments up to age 35.

In both longitudinal studies, those with notable STEM accomplishments had been involved in a richer and more robust collection of advanced precollegiate educational opportunities in STEM ("STEM doses") than the members of their cohorts with lower levels of STEM-related professional achievement. This finding holds for students of both sexes. The types of "STEM doses" noted in these studies include advanced placement (AP) and early college math and science courses, science or math project competitions, independent research projects, and writing articles within the disciplines. Of these mathematically inclined students, those who participated in more than the median number of science and math courses and activities during their K-12 school years were about twice as likely, by age 33, to have earned a doctorate, become tenured, or published in a STEM field than were students who participated in a lower-than-average number of such activities. The differences in achieving a STEM professional occupation or securing a STEM patent between the "low dose" and "high dose" students were evident but not as pronounced. Note, however, that these results are merely an association and do not imply a cause-and-effect relationship. For example, those with the most interest and abilities in STEM fields might self-select for the enrichment programs. Nevertheless, it does fit with the individual experiences of many members of this committee that early exposure to highly challenging material in the mathematical sciences had an impact on their career trajectories.

One means by which the mathematical sciences professional community contributes to efforts to attract and encourage precollege students is through Math Circles. Box 5-1 gives an overview of this mechanism, which has proved to be of real value in attracting and encouraging young people with strong talent in the mathematical sciences.

From 1988 to 1996, the National Science Foundation (NSF) sponsored a Young Scholars Program that supported summer enrichment activities for high school students who exhibited special talent in mathematics and science.[33] It was begun at a time when the United States was worried about the pipeline for scientists and engineers just as it worries now. By 1996, the NSF was "funding 114 summer programs that reached around 5,000 students annually [and about] 15% of the Young Scholars programs were in mathematics."[34] Some of the successful funding of mathematics programs through this mechanism included programs at Ohio State University, Boston University, and Hampshire College. The committee believes

---

[33] This description is drawn from Allyn Jackson, 1998, The demise of the Young Scholars Program. *Notices of the AMS*, March.

[34] Ibid.

BOX 5-1
Mathematical Circles: Teaching Students to Explore[a]

In 2006, an eighth-grade home-schooled student named Evan O'Dorney came to an evening meeting of the Berkeley Mathematics Circle with his mother. For an hour he listened to the director, Zvezdelina Stankova, talk about how to solve geometry problems with a technique called circle inversion. Then, during a 5-minute break, he went back to his mother and told her, "Mom, there are problems here I can't do!"

It's not something that O'Dorney has said very often in his life. By the time he graduated from high school, he had become as famous for academic excellence as any student can be. In 2007, he won the National Spelling Bee. From 2008 to 2010 he participated in the International Mathematics Olympiad (IMO) for the U.S. team three times, winning two silver medals and a gold. And in 2011 he won the Intel Science Talent Search with a mathematics project on continued fractions. President Barack Obama called O'Dorney personally to congratulate him after his IMO triumph, and the two met in person during the Intel finals.

It would be easy to say that a student as talented as O'Dorney probably would have achieved great things even without the Berkeley Math Circle. But that would miss the point. For 5 years, the mathematics circle gave him direction, inspiration, and advice. It put him in contact with university professors who could pose problems difficult enough to challenge him. (As a ninth-grader, he took a university course on linear algebra and found a solution to a previously unsolved problem.) By the time he was a high-school senior, he was experienced enough and confident enough to teach sessions of the Berkeley Mathematics Circle himself. The experience helped him develop the communication skills he needed to win the Intel Science Talent Search.

Not all students can be O'Dorneys, of course. But the math circle concept, imported from Eastern Europe, has begun to find fertile ground in the United States. The National Association of Math Circles now counts 97 active circles in 31 states, most of them based at universities and led by university professors. As is the case in Eastern Europe, math circles have become one of the most effective ways for professional mathematicians to make direct contact with precollege students. In math circles, students learn that there is mathematics beyond the school curriculum. And yes, they discover problems that might be too hard for them to solve. But that is exactly the kind of problem that a student like O'Dorney *wants* to work on. Gifted students are often completely turned off by the problems they see in their high-school classes, which for them are as about as challenging as a game of tic-tac-toe.

Dr. Stankova, who was then a postdoctoral fellow at the Mathematical Sciences Research Institute at Berkeley (she now teaches at Mills College in Oakland), began the Berkeley Math Circle in 1998, hoping to replicate the experience she had as a grade-school student in Bulgaria. In Bulgaria and throughout Eastern Europe, math circles are found in most grade schools and many high schools. Just as students with a talent for soccer might play on a school soccer team, students with a talent for mathematics go to a math circle. This does not mean that the

continued

BOX 5-1 Continued

school's regular math curriculum is insufficient or inadequate; it simply recognizes that some students want more.

Dr. Stankova was surprised that a similar system did not exist in the United States. (The first math circle in the United States was founded at Harvard by Robert and Ellen Kaplan in 1994; Stankova's was the second.) Originally the Berkeley Math Circle was intended as a demonstration for a program that would move into secondary schools.

But the United States turned out to be different from Eastern Europe in important ways. Here, very few secondary school teachers had the knowledge, the confidence, or the incentive to start a math circle and keep it going. This was different from the situation in Bulgaria, where schoolteachers were compensated for their work with math circles. Although some U.S. math circles have flourished without a university nearby (for example, the math circle in Payton, Illinois), most have depended on leadership from one or more university mathematicians. For example, the Los Angeles Math Circle has very close ties to the math department at UCLA.

Other differences showed up over time. With circles based at universities, logistics—getting kids to the meeting, and finding rooms for them to meet in—became more difficult. At present the Berkeley Math Circle, with more than 200 students, literally uses every seminar room available within the UC Berkeley math department on Tuesday nights. Most universities offer little or no support to the faculty who participate. Administrators do not always realize that the high-school students who attend the math circles are potential future star students at their universities. In fact, some of them are already taking courses at the university. Stankova has often had to alert UC Berkeley faculty members to expect a tenth-grader in their classes who will outshine the much older college students.

One part of the math circles philosophy has, fortunately, survived its transplantation from Eastern Europe to America. Math circles encourage open-ended exploration, a style of learning that is seldom possible in high-school curricula that are packed to the brim with mandatory topics. Problems in a math circle are defined as interesting questions that one does not know at the outset how to answer—the exact opposite of "exercises." They introduce students to topics that are almost never taught in high school: for example, circle inversion, complex numbers, continued fractions (the subject of O'Dorney's Intel project), cryptology, topology, and mathematical games like Nim and Chomp.

Many participants in math circles have gone on to success in scholastic math competitions, such as the USA Mathematical Olympiad (USAMO) and the IMO. For example, Gabriel Carroll, from the Berkeley Math Circle, earned a silver medal and two golds in the IMO, including a perfect score in 2001. He participated in the Intel Science Talent Search and finished third. As a graduate student in economics at MIT, Carroll proposed problems that were selected for both the 2009 and 2010 IMO events. Ironically, the latter problem was the only one that stumped O'Dorney.

But not all students are interested in competitions. Victoria Wood participated in the local Bay Area Math Olympiad but did not like having to solve problems in a limited time. She liked problems that required longer reflection (as real research

continued

BOX 5-1 Continued

problems almost always do). She started attending the Berkeley Math Circle at age 11, matriculated at UC Berkeley at age 13, and is now a graduate student with several patents to her name. Some math circles, such as the Kaplans' original math circle in Boston, deliberately avoid preparing students for math competitions. Others do provide preparation for competition, but it is far from being their main emphasis.

In 2006, the American Institute of Mathematics (AIM) began organizing math teachers' circles, designed specifically for middle-school teachers. After all, why should students have all the fun? By exposing teachers to open-ended learning, and encouraging them to view themselves as mathematicians, the organizers hope to have a trickle-down effect on thousands of students. At present, AIM lists 30 active teachers' circles in 19 states.

Despite their very promising start, it remains to be seen whether math circles will become a formal part of the American educational system or remain a poorly funded adjunct that depends on the passion and unpaid labor of volunteers. Clearly they have already provided an invaluable service to some of America's brightest youngsters. Conceivably, if teachers' circles take root, or if enough teachers come to observe math circles with their students, they could begin transforming American schools in a broader way, so that mathematical competence is expected and mathematical virtuosity is rewarded.

---

[a]The committee thanks Dana Mackenzie for drafting the text in this box.

---

that reviving this sort of program would contribute in exciting ways to the mathematical sciences (or STEM) pipeline.

> **Recommendation 5-5: The federal government should establish a national program to provide extended enrichment opportunities for students with unusual talent in the mathematical sciences. The program would fund activities to help those students develop their talents and enhance the likelihood of their pursuing careers in the mathematical sciences.**

In making this recommendation, the committee does not intend in any way to detract from the important goal of ensuring that every student has access to excellent teachers and training in the mathematical sciences. The goal of growing the mathematical sciences talent pool broadly is synergistic with the goal of attracting and preparing those with exceptional talent for high-impact careers in the mathematical sciences.

6

The Changing Academic Context

In addition to the trends discussed in Chapter 4, the mathematical sciences are also being affected by pressures on the academic environment. This chapter discusses emerging changes in, and pressures on, academe that appear likely to affect academic mathematical scientists. As noted in a recent report from the National Research Council, all of the normal funding streams of research universities are under stress:

> American research universities are facing critical challenges. First, their financial health is endangered as each of their major sources of revenue has been undermined or contested. Federal funding for research has flattened or declined; in the face of economic pressures and changing policy priorities, states are either unwilling or unable to continue support for their public research universities at world-class levels; endowments have deteriorated significantly in the recent recession; and tuition has risen beyond the reach of many American families. At the same time, research universities also face strong forces of change that present both challenges and opportunities: demographic shifts in the U.S. population, transformative technologies, changes in the organization and scale of research, a global intensification of research networks, and changing relationships between research universities and industry. In addition, U.S. universities face growing competition from their counterparts abroad, and the nation's global leadership in higher education, unassailable for a generation, is now threatened.[1]

---

[1] National Research Council, 2012, *Research Universities and the Future of America: Ten Breakthrough Actions Vital to Our Nation's Prosperity and Security*. The National Academies Press, Washington, D.C. pp. 3-4.

The mathematical sciences are likely to experience stresses and disruptions in the coming decade and a half, affecting both research and teaching. The business model of mathematical sciences departments will undergo major changes, owing to cost pressures, online course offerings, and so on. There may be less demand for lower-division teaching, but expanded opportunities for training students from other disciplines and people already in the workforce.[2] Mathematical scientists should work proactively—through funding agencies, university administrations, professional societies, and within their departments—to be ready for these changes.

Mathematical science departments, particularly those at large state universities, have a tradition of teaching service courses for nonmajors. These courses, especially the large lower-division ones, help to fund positions for mathematical scientists at all levels, but especially for junior faculty and graduate teaching assistants. The teaching of mathematical sciences, both to majors and nonmajors, justifies the positions of a substantial portion of those faculty members performing mathematical sciences research. But this business model is already changing, and it faces a number of challenges in the coming years. University education has become more expensive, straining family budgets severely and often leaving students with substantial debt when they graduate. The desire to reduce these costs is pushing students to take some of their lower-division studies at state and community colleges. It is also leading university administrations to hire a second tier of nonladder faculty with larger teaching loads, reduced expectations of research productivity, and lower salaries, or to implement a series of online courses that can be taught with less faculty involvement. New methods of teaching, particularly for introductory courses, may precipitate changes in the existing model. While these trends have been observed for a decade or more, financial concerns may be increasing pressure to shift more teaching responsibilities in these ways. The result could be a reduction in the number of faculty slots in many departments.

The pressure to economize is, if anything, increasing. In his 2012 State of the Union speech, President Obama said, "So let me put colleges and universities on notice: If you can't stop tuition from going up, the funding you get from taxpayers will go down. Higher education can't be a luxury—it is an economic imperative that every family in America should be able to afford." Three days later he unveiled "a financial aid overhaul that for the first time

---

[2] An analysis from the National Science Foundation (NSF) (Kelly Kang, 2012, "Graduate Enrollment in Science and Engineering Grew Substantially in the Past Decade but Slowed in 2010," *InfoBrief* from NSF's National Center for Science and Engineering Statistics, NSF 21-317, available at http://www.nsf.gov/statistics/ infbrief/nsf12317/nsf12317.pdf) found that overall graduate enrollment in science and engineering grew 35 percent from 2000 to 2010, to more than 550,000. As documented in Chapter 3 of this report, many science and engineering fields are increasingly reliant on the mathematical sciences.

would tie colleges' eligibility for campus-based aid programs—Perkins loans, work-study jobs and supplemental grants for low-income students—to the institutions' success in improving affordability and value for students."[3]

At the same time that these changes are taking place, there are countervailing opportunities. As discussed in Chapter 3, there is a broadening and overall expansion in the number of applications of the mathematical sciences. This increases the number of students who may be interested in courses within mathematical sciences departments, including some at the upper-division level. In addition, career paths in an expanding palette of areas come with an expectation of mid-career acquisition of new quantitative skills. Creating pathways for those already in the workplace to learn these new skill sets provides a major opportunity for mathematical sciences departments.

How mathematical sciences departments adapt to and manage these changes and opportunities will strongly affect the health of the profession and the quality of education offered by U.S. universities. The pace of change to the business model for education may well be similar in magnitude to that which currently roils the publishing industry. The mathematical sciences community needs to get out ahead of these potential changes and proactively make the most of its new opportunities.

Universities are also feeling other pressures that, directly or indirectly, could affect the state of the mathematical sciences in 2025. For example, many graduate students from overseas pay full tuition, so there is some incentive for universities to actively recruit them. In particular, self-funded master's students from abroad, or students seeking professional master's degrees, can be helpful to department finances, but will too many such students change the research environment?

Fiscal stresses on colleges and universities are also leading to the establishment of some for-profit educational institutions. This trend took root for continuing education, but it is now playing an increasing role in undergraduate education. It is difficult to say how widespread for-profit colleges and universities may become or how their presence might change the environment for the mathematical sciences, but it is a trend that mathematical scientists should monitor. In traditional settings, some educators are experimenting with lower-cost ways of providing education, such as Web-based courses that put much more burden on the students, thereby allowing individual professors to serve larger numbers of students. Mathematics and statistics, because they do not involve laboratory work, would appear to be promising targets for online delivery.

For example, the Math Emporium at Virginia Tech uses four untenured mathematics instructors to lead seven entry-level courses with enrollments

---

[3] Tamar Lewin, 2012, Obama plan links college aid with affordability, *New York Times*, January 27.

of between 200 and 2,000, for a total of 8,000 students per year, according to a 2012 article in the *Washington Post*.[4] According to that article, "Virginia Tech students pass introductory math courses at a higher rate now than 15 years ago, when the Emporium was built. And research has found the teaching model trims per-student expense by more than one-third, vital savings for public institutions with dwindling state support." It goes on to quote Carol Twigg, president of the nonprofit National Center for Academic Transformation, that the Emporium model has been adopted by about 100 colleges and community colleges.

In general, there is pressure to find less costly means of delivering classroom knowledge. An extreme scenario would be greater decoupling of teaching and research, with fewer universities focused on leading research. Movement in that direction would have a large impact on the mathematical sciences because the size of most mathematical science departments is driven by the teaching load. If teaching duties are offloaded to other mechanisms (community colleges, online learning, for-profit institutions), university mathematics and statistics departments may lose some critical mass. Such a reduction in service teaching could also weaken ties between mathematical scientists and other departments.

Some online courses with mathematical content have already proven to be tremendously popular, and this early attention will only increase the interest (by students and university administrations, at least) in experimenting with this modality. A 2012 article in the *New York Times*[5] pointed to the enormous number of people around the globe who enrolled in courses offered in the fall of 2011 by Stanford University: 160,000 students in 190 countries enrolled for a course in artificial intelligence, 104,000 for a course in machine learning, and 92,000 for an introductory database course. According to that article, other major universities, such as the Massachusetts Institute of Technology (MIT) and the Georgia Institute of Technology, are also beginning to offer "massive, open, online courses" or MOOCs. Other courses with mathematical content are offered through Coursera.org, which "is committed to making the best education in the world freely available to any person who seeks it."[6] As of October 11, 2012, the listings included the following:

- Model Thinking, from the University of Michigan;
- Introduction to Mathematical Thinking, from Stanford University;

---

[4] Daniel de Vise, 2012, At Virginia Tech, computers help solve a math class problem. *Washington Post*, April 22.

[5] Tamar Lewin, 2012, Instruction for masses knocks down campus walls. *New York Times*, March 4.

[6] From https://www.coursera.org/landing/hub.php.

- Algebra, from the University of California, Irvine;
- Calculus: Single Variable, from the University of Pennsylvania;
- Analytic Combinatorics, from Princeton University; and
- Machine Learning, from the University of Washington.[7]

More recently, Harvard and MIT announced a joint partnership called edX "to offer online learning to millions of people around the world. EdX will offer Harvard and MIT classes online for free."[8] The press release[9] accompanying that announcement notes that online students may receive "certificates of mastery" if they demonstrate adequate knowledge of the course material. It also states that "edX will release its learning platform as open-source software so it can be used by other universities and organizations that wish to host the platform themselves." The press release goes on to say that Harvard and MIT faculty will use data from edX "to research how students learn and how technologies can facilitate effective teaching both on-campus and online . . . [to study] which teaching methods and tools are most successful."

At the same time that mathematics and statistics departments are feeling these pressures, there is also the challenge noted at the beginning of Chapter 5: the belief in some circles that more lower-division mathematics should be taught by other departments. The 2012 report of the President's Council of Advisors on Science and Technology on STEM education at the undergraduate level recommended that this hypothesis be actively explored through a set of perhaps 200 experiments across the nation. As stated in Chapter 5, the committee agrees that the existing mathematics curriculum would benefit from a significant updating of both content and teaching techniques. There is a real chance that if mathematicians do not do this, others will, and that could exacerbate the erosion in mathematics service teaching that is likely to occur due to cost pressures.

Another important trend of concern to all STEM disciplines is that graduate enrollments from overseas are likely to go down over time as the quality of overseas universities improves, because employment opportunities now exist worldwide for mathematical sciences talent. Over half (52 percent in the 2009-2010 academic year) of the Ph.D. degrees awarded annually in the mathematical sciences by U.S. universities are to non-U.S. citizens.[10] Until now, a large fraction of them have continued their careers in the United States, and the nation has benefited greatly in recent decades

---

[7] Ibid.

[8] From http://www.edxonline.org/. Accessed May 10, 2012.

[9] Available at http://web.mit.edu/press/2012/mit-harvard-edx-announcement.html. Accessed May 2, 2012.

[10] R. Cleary, J.W. Maxwell, and C. Rose, 2010, Report on the 2009-2010 new doctoral recipients. *Notices of the AMS* 58(7):944-954.

because of its ability to attract such people, many of whom stay to contribute to U.S. science, technology, and business.

However, as economic and scientific conditions improve in other countries—especially in China and India—it may be more difficult to keep foreign-born graduates in the United States; already, other nations are aggressively recruiting talented individuals, especially those born there but who are now in the United States. Increasingly, there are reports of more Chinese graduate students electing to return to China after their Ph.D. work, and the opportunities for rewarding research careers in the mathematical sciences are improving in China and elsewhere overseas. Publication counts also suggest that other locations are increasingly productive in mathematics. From 1988 through 2003, the number of publications in mathematics worldwide increased by 40 percent—from 9,707 to 15,170—while the number of mathematics publications with at least one U.S. author increased by only 8 percent—from 4,301 to 4,651.[11] U.S. policies regarding work visas and immigration are an important factor here, too. A decline in the ability of the United States to attract and retain top international students will have a serious negative effect on U.S. graduate training and on the production of young mathematical scientists to meet the demand of U.S. academic institutions, industry, and government exactly at the time of increasing demand for such people.

To the extent possible, NSF policies should be aligned with the goals of continuing to attract top foreign talent to our shores and inducing talented foreigners, especially those who pass through our educational system, to choose to make their careers here. Policies that encourage the growth of the U.S.-born segment of the mathematical sciences talent pool should clearly continue, but they need to be supplemented by programs to attract and retain mathematical scientists from other countries, especially for graduate school and continuing as feasible into early careers. This goal leads directly to questions about immigration policies, which are, of course, beyond the control of NSF. Mathematical scientists who are concerned about the future vitality of our profession should recognize the important role played by immigration policies and perhaps weigh in on related political discussions.

One particular aspect deserves mention here in connection with the stresses on academic finances: The ratio of federal support to institutional support for graduate students in the mathematical sciences is very low relative to the same ratio for students of other sciences, as shown in Figure 6-1. The support model for graduate students in the mathematical sciences is overly reliant on teaching assistantships, which extends time to degree,

---

[11] Derek Hill, Alan I. Rapoport, Rolf F. Lehming, and Robert K. Bell, 2007, Changing U.S. output of scientific articles: 1988-2003. Report 07-320, Appendix Table 2. National Science Foundation, Division of Science Resources Statistics, Arlington, Va.

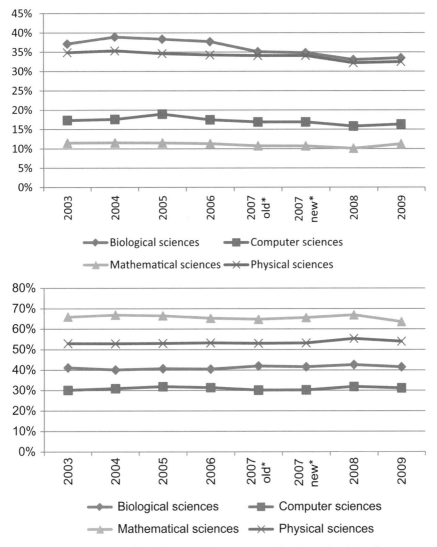

FIGURE 6-1 Fraction of graduate students supported (*above*) by federal programs (primarily for research assistantships) and (*below*) by their academic institutions (primarily for teaching assistantships). In 2007, Graduate Student Support (GSS)-eligible fields were reclassified, newly eligible fields were added, and the survey was redesigned to improve coverage and coding of GSS-eligible units. "2007 new" presents data as collected in 2007; "2007 old" reflects data as they would have been collected under 2006 methodology. SOURCE: National Science Foundation/ National Center for Science and Engineering Statistics, 2009, *NSF-NIH Survey of Graduate Students and Postdoctorates in Science and Engineering,* Table 38. Available at http://www.nsf.gov/statistics/nsf12300/content.cfm?pub_id=4118&id=2.

and is especially burdensome at a time when the amount that a graduate student in the mathematical sciences must learn is expanding. Overreliance on teaching assistantships is also worrisome because the changing business model for mathematics departments makes this source of support especially vulnerable to cutbacks, as discussed above. As a community, the mathematical sciences must be proactive in shifting this balance, because innovations in the delivery of the classes that support teaching assistants could erode that means of support much faster than the number of research assistantships could ramp up. A first step is for mathematical science researchers to be more aggressive in seeking research assistantships for their students, in recognition of the need for graduate students today to gain more research experience and to lessen departments' dependency on teaching assistantships.

One additional pressure of particular relevance to the mathematical sciences is the movement toward more multidisciplinarity in research, as emerging fields require mathematics and statistics expertise in order to move forward. At one extreme, this could lead to situations in which more mathematical scientists are members of the departments in which their work is applied, so that mathematics or statistics departments lose critical mass. If the mathematical sciences were to become dispersed in this way, the coherence and unity of the field would be threatened. Forging links to other departments will help in advancing this process. Such links would include cross-listing of courses, collaboration with other departments in planning courses, and having cross-disciplinary postdoctoral students and courtesy appointments. Creating appropriate methods to evaluate those engaged in interdisciplinary research is overdue.

The multitude of existing configurations of mathematical sciences departments at academic institutions often reflect the particular history at each institution rather than what is optimal. In view of the changing academic environment there is an opportunity to reconsider such arrangements and departmental divisions, in order to enhance the cohesiveness of the mathematical sciences and enable intradisciplinary and cross-disciplinary research and educational collaborations.

> **Recommendation 6-1: Academic departments in mathematics and statistics should begin the process of rethinking and adapting their programs to keep pace with the evolving academic environment, and be sure they have a seat at the table as online content and other innovations in the delivery of mathematical science coursework are created. The professional societies have important roles to play in mobilizing the community in these matters, through mechanisms such as opinion articles, online discussion groups, policy monitoring, and conferences.**

Appendixes

Appendix A

Past Strategic Studies

The first strategic examination of the mathematical sciences, *Renewing U.S. Mathematics: Critical Resource for the Future* (known as the "David report"),[1] found worrisome trends. Although the mathematical sciences were producing excellent and valuable research, the number of young people entering the profession had declined, threatening a contraction in the size of the mathematical science research enterprise. The report documented an erosion in federal support for mathematical sciences research that had taken place over more than a decade. The result was an imbalance between research in the mathematical sciences and research in the physical sciences and engineering, which depend on mathematical and statistical tools. For example, the report cited 1980 figures on the number of faculty members in chemistry, physics, and mathematical sciences (three fields with very similar numbers of faculty members) who received federal research funding. Approximately 3,300 chemists and 3,300 physicists received federal research funding in 1980, compared with just 2,300 mathematical scientists.[2] The study estimated the number of graduate student research associateships and postdoctoral research positions that would be needed for a healthy pipeline of new researchers and, from that, argued for a doubling of federal funding for mathematical sciences research.

The David report (named after its chair, former Presidential Science Advisor Edward David) led to striking increases in federal funding for math-

---

[1] National Research Council (NRC), 1984, *Renewing Mathematics: Critical Resource for the Future*. The National Academies Press, Washington, D.C.

[2] Ibid., p. 5.

ematical sciences for a few years, partially restoring balance, though the increases shrank later in the 1980s before the doubling goal was reached. The report also stirred a good deal of discussion within the community and led to greater involvement of mathematical sciences in discussions about federal science policy. According to the subsequent "David II report,"[3] members of the mathematical sciences community had "shown a growing awareness of the problems confronting their discipline and increased interest in dealing with the problems, particularly in regard to communication with the public and government agencies and involvement in education."

Because the imbalance in federal funding was only partially remedied as a result of the David report—federal funding for mathematical sciences research increased by 34 percent, not 100 percent—the funding agencies that support the mathematical sciences decided in 1989 to commission the David II report to assess progress and recommend further steps to strengthen the enterprise. That report found that federal support for graduate and postdoctoral students had increased substantially between 1984 and 1989—by 61 percent and 42 percent, respectively—and some aspects of infrastructure, such as computing facilities and research institutes, had been upgraded. But overall, the David II report found that the foundations of the research enterprise continued to be "as shaky now as in 1984."[4] It reiterated the calls of the first David report and recommended continued work toward doubling of federal support. It also recommended improvements to the career path in the mathematical sciences, through increases in the number of researchers, postdoctoral research positions, and graduate research associateships, all of which indeed did grow during the 1990s. It is not clear, though, that those steps reduced the degree to which U.S. students from high school onward leave the mathematical sciences pipeline. The David II report also asserted specifically that "recruitment of women and minorities into the mathematical sciences is a high priority," but it did not propose concrete steps to improve this. In general, the early 1990s was not a favorable time for a renewed push for federal funding, and it is not clear whether the David II report had much impact in that area.

In 1997, the National Science Foundation (NSF's) Division of Mathematical Sciences (DMS) organized the Senior Assessment Panel for the International Assessment of the U.S. Mathematical Sciences. The study was meant to evaluate how well DMS was supporting NSF's strategic goals with respect to the mathematical sciences—which included "enabl[ing] the United States to uphold a position of world leadership in all aspects of . . . mathematics . . . promot[ing] the discovery, integration, dissemination, and

---

[3] NRC, 1990, *Renewing U.S. Mathematics: A Plan for the 1990s*. National Academy Press, Washington, D.C., p. 3.

[4] Ibid.

employment of new knowledge in the service of society; and achiev[ing] excellence in U.S. science, mathematics, engineering and technology education at all levels.[5] The panel was chaired by Lt. Gen. William E. Odom, former head of the National Security Agency.

The Executive Summary of the Odom report reaches conclusions and makes recommendations:

> The modern world increasingly depends on the mathematical sciences in areas ranging from national security and medical technology to computer software, telecommunications, and investment policy. More and more American workers, from the boardroom to the assembly line, cannot do their jobs without mathematical skills. Without strong resources in the mathematical sciences, America will not retain its pre-eminence in industry and commerce.
>
> At this moment, the U.S. enjoys a position of world leadership in the mathematical sciences. But this position is fragile. It depends very substantially on immigrants who had their mathematical training elsewhere and in particular on the massive flow of experts from the former Communist bloc. . . . Young Americans do not see careers in the mathematical sciences as attractive. Funding for graduate study is scarce and ungenerous, especially when compared to funding for other sciences and with what happens in Western Europe. Further, it takes too long to obtain a doctorate because of the distractions of excessive teaching. Students wrongly believe that jobs that call for mathematical training are scarce and poorly paid. Weaknesses in K-12 mathematics education undermine the capabilities of the U.S. workforce.
>
> Based on present trends, it is unlikely that the U.S. will be able to maintain its world leadership in the mathematical sciences. It is, however, essential for the U.S. to remain the world leader in critical subfields, and to maintain enough strength in all subfields to be able to take full advantage of mathematics developed elsewhere. Without remedial action by the universities and [the NSF], the U.S. will not remain strong in mathematics: there will not be enough excellent U.S.-trained mathematicians, nor will it be practicable to import enough experts from elsewhere, to fill the Nation's needs. . . .
>
> We recommend that [the NSF] encourage programs that:
>
> - Broaden graduate and undergraduate education in the mathematical sciences. Provide support for full time graduate students in the mathematical sciences comparable with the other sciences.
> - Provide increased opportunity for postdoctoral study for those who wish to become academic researchers as a means to broaden and strengthen their training as professional mathematicians.

---

[5] National Science Foundation, 1998, *Report of the Senior Assessment Panel for the International Assessment of the U.S. Mathematical Sciences.* NSF, Arlington, Va., p. ii.

- Encourage and foster interactions between university-based mathematical scientists and users of mathematics in industry, government, and other disciplines in universities.
- Maintain and enhance the historical strength of the mathematical sciences in its academic setting as an intellectual endeavor and as a foundation for applications, sustaining the United States position of world leadership.[6]

The release of the Odom report was followed by strong increases in funding for NSF/DMS, with the division's budget nearly doubling from FY2000 to FY2004, when it reached $200 million per year. The NSF director at that time, Rita Colwell, was very supportive of the mathematical sciences and encouraged a number of new initiatives, including partnerships between DMS and other NSF units. In addition, DMS began programs aimed at improving career preparation for the future mathematical sciences workforce and substantially broadened the portfolio of mathematical sciences research institutes.

---

[6] Op. cit., pp. 1-2.

Appendix B

Meeting Agendas and Other Inputs to the Study

MEETING 1
SEPTEMBER 20 AND 21, 2010
WASHINGTON, D.C.

Discussion of study goals with sponsor	**Sastry Pantula**, National Science Foundation (NSF) **Deborah Lockhart**, NSF
Discussion of study goals with major professional societies	**James Crowley**, executive director, Society for Industrial and Applied Mathematics (SIAM) **Tina Straley**, executive director, Mathematical Association of America (MAA) **Ron Wasserstein**, executive director, American Statistical Association (ASA) **Donald McClure**, executive director, American Mathematical Society (AMS)
What changes and stresses are affecting the research enterprise?	**William E. Kirwan**, mathematician and chancellor of the University System of Maryland **C. Judson King**, former Berkeley provost and director of its Center for Studies in Higher Education

Possible models for our study	**Philip Bucksbaum**, Stanford University, co-chair of *Controlling the Quantum World: The Science of Atoms, Molecules, and Photons* (2007) **Donald Shapero**, director of the National Research Council's Board on Physics and Astronomy
Funding for mathematical sciences research	**Sastry Pantula**, NSF **Deborah Lockhart**, NSF **Walter Polansky**, DOE **Wen Masters**, DoD **Charles Toll**, National Security Agency (NSA) **David Eisenbud**, Simons Foundation **James Crowley**, executive director, SIAM (discussing industry research)
Major advances in recent years that illustrate new opportunities and future directions	**James Carlson**, president, Clay Mathematics Institute

MEETING 2
DECEMBER 4 AND 5, 2010
IRVINE, CALIFORNIA

The changing university environment	**Hal S. Stern**, dean and professor of statistics, School of Information and Computer Sciences, University of California, Irvine
The demand for mathematical science skills in biology	**Terrence Sejnowski**, Salk Institute for Biological Studies, University of California, San Diego
The demand for mathematical science skills at the NSA	**Alfred Hales**, UCLA (ret.), former director of the Institute for Defense Analyses' Center for Communications Research–La Jolla
Recent changes for the mathematical sciences in China	**S.-T. Yau**, Harvard University

The demand for mathematical science skills at DreamWorks Studios

Nafees Bin Zafar, DreamWorks

The demand for mathematical science skills in the financial sector

James Simons, Renaissance Technologies

The demand for mathematical science skills at Microsoft, and experience establishing a research center in Beijing

Harry Shum, Microsoft

The demand for mathematical science skills at IBM

Brenda Dietrich, vice president for Business Analytics and Mathematical Sciences at the IBM T. J. Watson Research Center

MEETING 3
MAY 12 AND 13, 2011
CHICAGO, ILLINOIS

Stresses and opportunities for the mathematical sciences

Robert Fefferman, dean of physical sciences, University of Chicago
Robert Zimmer, President, University of Chicago

What are major opportunities for the mathematical sciences, steps needed to realize them, and stresses affecting the profession over the coming years?

Yali Amit, University of Chicago, Statistics Department
Peter Constantin, University of Chicago, Mathematics Department
Kam Tsui, University of Wisconsin, Statistics Department
Douglas Simpson, University of Illinois at Urbana-Champaign, Statistics Department
Bryna Kra, Northwestern University, Mathematics Department
Lawrence Ein, University of Illinois at Chicago, Mathematics, Statistics, and Computer Science Department
Shi Jin, University of Wisconsin, Mathematics Department
William Cleveland, Purdue University, Statistics Department

INPUTS FROM LEADING MATHEMATICAL
SCIENCE RESEARCHERS

Between March 1 and May 2, 2011, the committee held a series of conference calls with the following leading researchers in the mathematical sciences:

Emery Brown, Massachusetts General Hospital
Ronald Coifman, Yale University
David Donoho, Stanford University
Cynthia Dwork, Microsoft Research
Charles Fefferman, Princeton University
Jill Mesirov, Broad Institute
Assaf Naor, New York University
Martin Nowak, Harvard University
Adrian Raftery, University of Washington
Terence Tao, University of California, Los Angeles
Richard Taylor, Harvard University

The purpose of these calls was to identify important trends and opportunities for the discipline, drawing on the diverse perspectives of research frontiers, and also to discuss any concerns these experts have about the future. The calls were very helpful, and some of the observations presented contributed to Chapters 3 and 5. Insights from these conference calls helped the committee select the recent advances that it highlighted in Chapter 2, and they also contributed to the identification of trends discussed in Chapter 4.

Appendix C

Basic Data about the
U.S. Mathematical Sciences

CURRENT AND RECENT FUNDING FOR
THE MATHEMATICAL SCIENCES

In order to provide a picture of the financial support for the mathematical sciences, this appendix gives an overview of federal funding and two private funding sources. Much of this funding supports not only research per se but also most of the research associateships and postdoctoral fellowships that prepare the next generation of researchers. A fraction of the federal funding supports workshops, research institutes, and other mechanisms that enable sharing of results and community interactions. Additional research funding for academic researchers is available from a variety of sources—universities, states, foundations, industry, and programs that primarily fund other science and engineering disciplines. In addition, a good deal of mathematical sciences research is carried out and supported in industry and in government laboratories. Researchers who perform mathematical sciences work in private corporations may not be labeled as "mathematician" or "statistician," so it is difficult to characterize the magnitude of that component of the mathematical sciences research enterprise. This report does not attempt to do so, but estimates do exist.[1]

Federal Funding for the Mathematical Sciences

Federal extramural funding for the mathematical sciences has increased in recent years. Four main governmental agencies and their suborganiza-

---

[1] Chapter 4 presents some information about the mathematical sciences in industry.

163

tions typically fund the mathematical sciences through extramural grants (Table C-1):

- National Science Foundation (NSF),
- Department of Defense (DOD),
  —Air Force Office of Scientific Research (AFOSR),
  —Army Research Office (ARO),
  —Defense Advanced Research Projects Agency (DARPA),
  —National Security Agency (NSA),
  —Office of Naval Research (ONR),
- National Institutes of Health (NIH),
  —National Institute of General Medical Sciences (NIGMS),
  —National Institute of Biomedical Imaging and Bioengineering (NIBIB), and
- Department of Energy (DOE),
  —Two programs of the Office of Advanced Scientific Computer Research (ASCR): Applied Mathematics and Scientific Discovery Through Advanced Computing (SciDAC)

NSF is consistently the largest single supporter of the mathematical sciences, and it is the sole federal agency that devotes a significant amount of funding to areas of core mathematics. (NSA's extramural program is focused on core mathematics, but it is quite small.)

Private Sector Mathematical Sciences Funding

The Simons Foundation is a relatively new source of funding for the mathematical sciences and is becoming a major source of support. The Simons Foundation Program for Mathematics and the Physical Sciences focuses on "the theoretical sciences radiating from Mathematics: in particular, the fields of Mathematics, Theoretical Computer Science and Theoretical Physics."[2] In 2009, the Simons Foundation launched a program to provide an estimated $40 million annually for research in mathematics and theoretical aspects of physical science that relate to mathematics. Much of this initial money went to fund 68 postdoctoral positions at 46 universities.

The Simons Foundation will also be funding 40 U.S. and Canadian academic researchers every year as part of the Simons Fellows program, which is intended to increase the opportunity for research leaves from classroom teaching and academic administration and to extend sabbatical leaves to last for a full academic year. At a smaller scale, individual grants of

---

[2] The Simons Foundation, "Mathematics and the Physical Sciences." Available at https://simonsfoundation.org/mathematics-physical-sciences. Accessed July 12, 2011.

TABLE C-1 Federal Funding for the Mathematical Sciences (millions of dollars)

	2005	2006	2007	2008	2009	ARRA[a]	2010	2011	2012 Estimate
NSF									
DMS	200	200	206	212	225	97	245	240	238
DOD									
AFOSR	30	32	35	37	45	0	52	58	47
ARO	10	14	14	12	13	0	12	16	16
DARPA	19	16	26	19	21	0	12	16	28
NSA	4	4	4	4	4	0	7	6	6
ONR	14	13	14	14	23	0	20	22	24
Total DOD	77	79	93	85	104	0	103	118	121
DOE									
Applied Math	30	32	33	32	45	0	44	46	46
SciDAC	0	3	42	54	60	0	50	53	44
Total DOE	30	36	75	86	105	0	94	99	90
NIH									
NIGMS	35	38	45	45	47	0	50	—	—
NIBIB	38	39	38	38	38	0	39	—	—
Total NIH	73	77	83	83	85	0	89	—	—
Total	380	391	456	466	519	97	531	457	449

NOTE: Budget information is approximate and has been derived from agency documents and AMS staff conversations with agency program managers and representatives. According to the author of the AMS reports, Samuel M. Rankin, III, in a personal communication on October 31, 2011, the amounts shown here for NIH are apparently dominated by intramural funding within NIH and therefore overstate the funds available for the broader community, while the totals shown for DOE include a large amount of funding awarded to researchers at DOE national laboratories. Other amounts in this table are extramural funds. This compilation does not capture some mathematical sciences research carried out in the Department of Commerce at the Census Bureau and the National Institute of Technology.

[a]American Recovery and Reinvestment Act of 2009.

SOURCE: Samuel M. Rankin, III, Mathematical sciences in the FY2011 budget. *Notices of the AMS* 57(8):988-991; ———, Mathematical sciences in the FY2010 budget. *Notices of the AMS* 56(8): 1285-1288; ———, Mathematical sciences in the FY2009 budget. *Notices of the AMS* 55(7): 809-812; ———, Mathematical sciences in the FY2008 budget. *Notices of the AMS* 54(7): 872-875; and ———, Mathematical sciences in the FY2007 budget, *Notices of the AMS* 53(6): 682-685. Data for 2012 and revised 2010 and 2011 data from personal communication with Samuel M. Rankin III on October 3, 2012.

no more than $7,000 are available specifically to defray expenses (such as travel) associated with collaborations. Other mathematical funding comes in the form of Math+X grants, which offer matching endowment grants (up to $1.5 million) to universities to create new tenured Chairs to be shared equally between a department of mathematics and another science or engineering department. The Math+X grants also support one postdoctoral researcher and two graduate students (totaling up to $325,000 annually). Finally, the foundation is funding a new institute for the theory of computing, supported at $6 million annually for up to 10 years.

The Clay Mathematics Institute (CMI), another privately funded organization, aims to stimulate and disseminate mathematics research. CMI supports individual research mathematicians at various stages in their careers and organizes conferences, workshops, and an annual summer school.[3]

Finally, the American Institute of Mathematics (AIM) is partially funded by private funds from the Fry Foundation, with additional funding from NSF. The goal of AIM is to expand the frontiers of mathematical knowledge through focused research projects, sponsored conferences, and the development of an online mathematics library.[4]

SOURCES OF FEDERAL FINANCIAL SUPPORT

National Science Foundation

The NSF is the primary federal funder of mathematical sciences research and the only one that provides significant extramural support for core fields. The focal point in NSF for the mathematical sciences is the Division of Mathematical Sciences (DMS), which supports the following programs:

- Algebra and Number Theory,
- Analysis,
- Applied Mathematics,
- Combinatorials,
- Computational Mathematics,
- Foundations,
- Geometric Analysis,
- Mathematical Biology,
- Probability and Statistics, and
- Topology.

---

[3] The Clay Mathematics Institute, "About the Clay Mathematics Institute." Available at http://www.claymath.org/about/. Accessed August 18, 2011.

[4] American Institute of Mathematics, "About AIM." Available at http://www.aimath.org/about/. Accessed August 18, 2011.

Other DMS funding opportunities include special research programs, training programs (such as research experiences for undergraduates, research training groups, and postdoctoral research fellowships), career development programs, and institutes.

The fraction of research proposals to DMS that received funding was 35 percent in 2007, 31 percent in 2008, and 37 percent in 2009. DMS received approximately 2,200 proposals per year between 2005 and 2008, with a bump up to 2,300 in 2009 when funding from the federal stimulus program (ARRA) supplemented NSF funds. DMS made approximately 680 awards each year in 2005, 2006, and 2007. In 2008, the total number of awards was 770, and in 2009 it was 840. The median award size in 2008 was $61,000.

Department of Defense

The Department of Defense (DOD) provides extramural funding for the mathematical sciences primarily through five organizations: AFOSR, ARO, DARPA, NSA, and ONR.[5] In aggregate, DOD is the second-largest federal funder of extramural mathematical sciences research. It also supports in-house R&D in the mathematical sciences, especially at the NSA, the Air Force Research Laboratory (AFRL), and Naval Research Laboratory (NRL). In particular, NSA is often said to be the largest employer of mathematicians in the United States. Details of these in-house research programs are not readily available, and they are not always clustered according to academic disciplines, so the following discussion covers only the extramural programs.

AFOSR accounts for the largest continued funding within DOD, as seen in Table C-1. The majority of AFOSR support for the mathematical sciences comes from the Mathematics, Information, and Life Sciences directorate. Mathematical science areas of particular interest to this directorate are collective behavior and sociocultural modeling; complex networks; computational mathematics; dynamics and control; science of information, computation, and fusion; information operations and security; mathematical modeling of cognition and decision; optimization and discrete mathematics; robust computational intelligence; and systems and software.

---

[5] Many arms of the DOD make use of advances in the mathematical sciences. For example, the Army Research Laboratory and the Air Force Research Laboratory have some collaborations (including funding) with mathematical science researchers, the National Geospatial-Intelligence Agency supports some academic research in the mathematical sciences, and several offices in the Pentagon are strongly connected with operations research. But, generally speaking, these sources of support are not formalized in ongoing extramural programs, and they are not captured in Table C-1.

ARO's extramural program for the mathematical sciences is modest by comparison. It funds research in four areas: probability and statistics, modeling of complex systems, numerical analysis, and biomathematics.

The majority of ONR research in the mathematical sciences comes from the Command, Control, Communications, Computers, Intelligence, Surveillance, and Reconnaissance (C4ISR) Department. This department is partitioned into three divisions: Mathematics, Computers, and Information Research; Electronics, Sensors, and Network Research; and Applications and Transitions. The first of these divisions focuses on research in applied computational analysis, command and control, image analysis and understanding, data analysis and understanding, information integration, intelligent and autonomous systems, mathematical optimization, signal processing, and software and computing systems. The Electronics, Sensors, and Network Research Division focuses on research in communications and networking analysis, signal processing, and a number of nonmathematical areas. The Applications and Transitions Division broadly focuses on programs in surface and aerospace surveillance, communications, and electronic combat.

Much of the funding for mathematical sciences within DARPA comes from the Defense Sciences Office (DSO). DSO's mathematics program has an applied and computational mathematics component, which includes applications such as signal and image processing, biology, materials, sensing, and design of complex systems. The mathematics program also has a fundamental mathematics component, which focuses on exploring selected core areas with the potential for being relevant to future applications. These fundamental areas include topological and geometric methods, extracting knowledge from data, and new approaches to connecting key areas of mathematics.

NSA has a very large in-house program in mathematical sciences research. Its extramural program is small but important because it is one of the few non-NSF sources of funding for areas of core mathematics, offering grants for unclassified research in algebra, number theory, discrete mathematics, probability, and statistics. There is also a research grants program, which offers funding to beginning, midlevel, and senior researchers. In addition, MSP funds conferences, workshops, special situation proposals, and sabbatical programs for mathematicians, statisticians, and computer scientists.

Department of Energy

The Applied Mathematics program at DOE's Office of Advanced Scientific Computing Research (ASCR) supports basic research leading to fundamental mathematical advances and computational breakthroughs relevant to DOE missions. Applied Mathematics research supports efforts to develop robust mathematical models, algorithms, and numerical software for enabling predictive scientific simulations of DOE-relevant complex systems. Research includes numerical methods for solving ordinary and partial differential equations, multiscale and multiphysics modeling, analysis and simulation, numerical methods for solving large systems of linear and nonlinear equations, optimization, uncertainty quantification, and analysis of large-scale data. The program addresses foundational, algorithmic, and extreme-scale mathematical challenges. Currently, approximately two-thirds of the funding supports DOE national laboratory researchers and one-third supports academic and industry researchers.

In addition, ASCR's SciDAC program supports development of mathematical methodologies, algorithms, libraries, and software for achieving portability and interoperability to accelerate the use of high-performance computing for DOE science.

National Institutes of Health

Several of the 27 institutes and centers (ICs) that comprise the NIH support the development of methodology and novel applications in mathematics, computing, and statistics applied to biological, biomedical, and behavioral research. Institutes that have shown particular interest in these areas are the National Institute of General Medical Sciences (NIGMS), the National Institute of Biomedical Imaging and Biomedical Engineering (NIBIB), the National Cancer Institute (NCI), and the National Human Genome Research Institute (NHGRI). NIGMS, in partnership with NSF/DMS, administers a Math-Biology Initiative that helps to make opportunities available to a broad range of the mathematical sciences community. The NIBIB focuses on applications of importance to imaging technologies and bioengineering. Only the extramural funding from NIGMS and NIBIB is captured in Table C-1.

Several NIH institutes, such as NCI, conduct significant amounts of in-house research in biostatistics, scientific computing, and computational biology, among other areas. These intramural programs are not captured in Table C-1. In addition, some other parts of the Department of Health and Human Services (NIH's parent organization), such as the Agency for Healthcare Research and Quality, support some extramural research in the mathematical sciences.

BASIC DATA ABOUT THE MATHEMATICAL SCIENCES POSTSECONDARY PIPELINE

The number of undergraduate degrees granted by mathematics and statistics departments in the United States annually over the period

2006-2010 has been roughly 24,000.[6] In 2010, 43 percent of these degrees were awarded to women and nearly half were from departments granting a baccalaureate degree only.

Master's degrees in the mathematical sciences are similarly tracked. Between 2006 and 2010, more than 4,000 such degrees were granted, with approximately 40 percent going to females.[7] Doctoral degrees in mathematics are on the rise, reaching 1,653 new Ph.D.s for 2010-2011. Figure C-1 gives a breakdown of the 2010-2011 degrees by department type.[8] Figure C-2 shows the trends in annual data by department type for the 2001-2002 to 2010-2011 academic years. Of the total doctorate degrees granted in 2010-2011, 49 percent of the recipients were U.S. citizens and 32 percent of the recipients were female. Figure C-3 gives some information about employment for new Ph.D. recipients from academic year 2010-2011. While most new Ph.D.s enter academic positions (including postdoctoral study at universities and research institutes), some 19 percent reported employment with government, business, or industry.

Finally, Figures C-4, C-5, and C-6 present basic data about the ethnic composition of mathematical science students at various postsecondary levels.

---

[6] American Mathematical Society, 2010, *Annual Survey of the Mathematical Sciences (AMS-ASA-IMS-MAA-SIAM)*; Supplementary Tables UD.1 and UD.2. Available at http://www.ams.org/profession/data/annual-survey/2010Survey-DepartmentalProfile-Supp-TableUD1-2.pdf .

[7] Ibid., *Section on Master's Degrees Awarded.* Supplementary Tables MD.1 and MD.2.

[8] Richard Cleary, James W. Maxwell, and Colleen Rose, 2012, Report on the 2010-2011 new doctoral recipients. *Notices of the AMS,* 58(8):1083-1093. Available at http://www.ams.org/notices/201208/rtx120801083p.pdf.

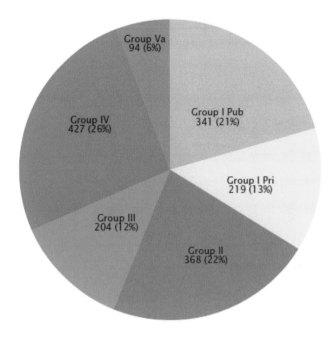

FIGURE C-1 Number and percentage of doctoral degrees awarded by department grouping in the mathematical sciences (1,653 total; between July 1, 2010, and June 30, 2011). Group I is composed of departments with scores in the 3.00-5.00 range according to rankings from the National Research Council. Group I Public and Group I Private are Group I departments at public institutions and private institutions, respectively. Group II is composed of 56 departments with scores in the 2.00-2.99 range. Group III contains the remaining U.S. departments reporting a doctoral program, including a number of departments not included in the 1995 ranking of program faculty. Group IV contains U.S. departments (or programs) of statistics, biostatistics, and biometrics reporting a doctoral program. Group Va consists of departments of applied mathematics/applied science. SOURCE: R. Cleary, J.W. Maxwell, and C. Rose, 2012, Report on the 2010-2011 new doctoral recipients. *Notices of the AMS* 59(8):1083-1093, Figure A.1. Available at http://www.ams.org/notices/201208/rtx20801083p.pdf.

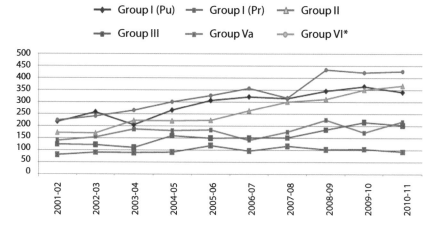

* The increase shown in Group VI is due in part to the increase in response rate

FIGURE C-2 Number and percentage of doctoral degrees awarded by department grouping in the mathematical sciences. The increase shown in Group IV is due in part to the increase in response rate. SOURCE: R. Cleary, J.W. Maxwell, and C. Rose, 2012, Report on the 2010-2011 new doctoral recipients. *Notices of the AMS* 59(8):1083-1093, Figure A.2. Available at http://www.ams.org/profession/data/annual-survey/2011Survey-NewDoctorates-Report.pdf.

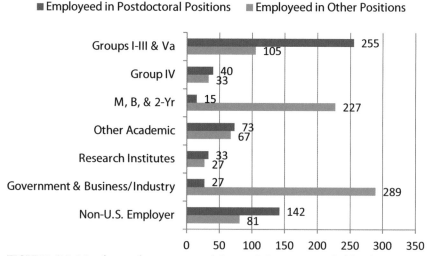

FIGURE C-3 Number and percentage of doctoral degrees awarded by department grouping in the mathematical sciences (1,653 total; between July 1, 2010, and June 30, 2011). SOURCE: R. Cleary, J.W. Maxwell, and C. Rose, 2012, Report on the 2010-2011 New Doctoral Recipients. *Notices of the AMS* 59(8):1083-1093, Figure E.5. Available at http://www.ams.org/notices/201208/rtx120801083p.pdf.

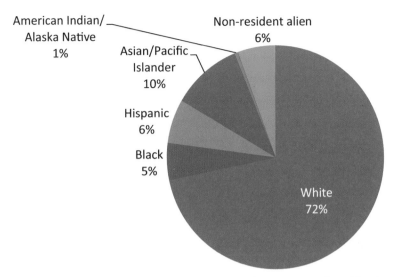

FIGURE C-4 Bachelor's degrees in mathematics and statistics conferred by degree-granting institutions, by race/ethnicity: 2010-2011. SOURCE: National Center for Education Statistics (NCES). Available at http://nces.ed.gov/programs/digest/ d12/ tables/dt12_301.asp, Table 300.

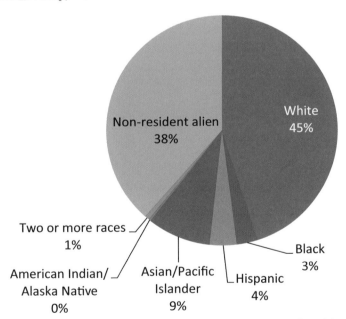

FIGURE C-5 Master's degrees in mathematics and statistics conferred by degree-granting institutions, by race/ethnicity, 2010-2011, SOURCE: Available at http:// nces.ed.gov/programs/digest/d12/tables/dt12_304.asp, Table 304.

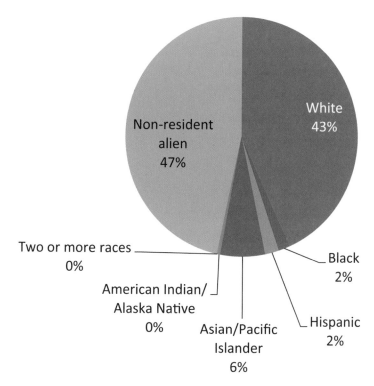

FIGURE C-6 Doctorates in mathematics and statistics conferred by degree-granting institutions, by race/ethnicity, 2010-2011. SOURCE: Available at http://nces.ed.gov/programs/digest/d12/tables/dt12_307.asp,Table 307.

Appendix D

Examples of the Mathematical Sciences Support of Science and Engineering

ILLUSTRATIVE DEMANDS FROM ASTRONOMY AND PHYSICS

The mathematical sciences continue to encounter productive challenges from physics, and there may be new opportunities in the coming years. For example, there are still open mathematical problems stemming from general relativity and from the mathematical descriptions of black holes and their rotation. The Laser Interferometer Gravitational-Wave Observatory (LIGO) project might produce data to stimulate these directions. In another direction, recent mathematical advances in understanding convergence properties of the Boltzmann equation might open the door to progress on some fundamental issues.

For some specific demands, the committee examined the 2003 NRC report *The Sun to the Earth—and Beyond: A Decadal Research Strategy in Solar and Space Physics*. This report poses several major challenges for that field that in turn pose associated challenges for the mathematical sciences. For example,

> Challenge 1: Understanding the structure and dynamics of the Sun's interior, the generation of solar magnetic fields, the origin of the solar cycle, the causes of solar activity, and the structure and dynamics of the corona.[1]

This challenge will require advances in multiscale methods and complex simulations involving turbulence.

---

[1] National Research Council, 2003. *The Sun to the Earth—and Beyond: A Decadal Research Strategy in Solar and Space Physics*. The National Academies Press, Washington, D.C., p. 2.

Later, that same report identified the following additional challenges that can only be addressed through related advances in the mathematical sciences:

> In the coming decade, the deployment of clusters of satellites and large arrays of ground-based instruments will provide a wealth of data over a very broad range of spatial scales. Theory and computational models will play a central role, hand in hand with data analysis, in integrating these data into first-principles models of plasma behavior. . . . The Coupling Complexity research initiative will address multiprocess coupling, non-linearity, and multiscale and multiregional feedback in space physics. The program advocates both the development of coupled global models and the synergistic investigation of well-chosen, distinct theoretical problems. For major advances to be made in understanding coupling complexity in space physics, sophisticated computational tools, fundamental theoretical analysis, and state-of-the-art data analysis must all be integrated under a single umbrella program.[2]
>
> The coming decade will see the availability of large space physics databases that will have to be integrated into physics-based numerical models. . . . The solar and space physics community has not until recently had to address the issue of data assimilation as seriously as have the meteorologists. However, this situation is changing rapidly, particularly in the ionospheric arena.[3]

Another example comes from the 2008 NRC report *The Potential Impact of High-End Capability Computing on Four Illustrative Fields of Science and Engineering*. This report identified the major research challenges in four disparate fields and the subset that depend on advances in computing. Those advances in computing are innately tied to research in the mathematical sciences. In the case of astrophysics, the report identified the following essential needs:

- N-*body codes*. Required to investigate the dynamics of collisionless dark matter, or to study stellar or planetary dynamics. The mathematical model is a set of first-order ODEs for each particle, with acceleration computed from the gravitational interaction of each particle with all the others. Integrating particle orbits requires standard methods for ODEs, with variable time stepping for close encounters. For the gravitational acceleration (the major computational challenge), direct summation, tree algorithms, and grid-based methods are all used to compute the gravitational potential from Poisson's equations.

---

[2] Ibid., pp. 64-66.
[3] Ibid., pp. 89-90.

- *PIC codes*. Required to study the dynamics of weakly collisional, dilute plasmas. The mathematical model consists of the relativistic equations of motion for particles, plus Maxwell's equations for the electric and magnetic fields they induce (a set of coupled first-order PDEs). Standard techniques are based on particle-in-cell (PIC) algorithms, in which Maxwell's equations are solved on a grid using finite-difference methods and the particle motion is calculated by standard ODE integrators.
- *Fluid dynamics*. Required for strongly collisional plasmas. The mathematical model comprises the standard equations of compressible fluid dynamics (the Euler equations, a set of hyperbolic PDEs), supplemented by Poisson's equation for self-gravity (an elliptic PDE), Maxwell's equation for magnetic fields (an additional set of hyperbolic PDEs), and the radiative transfer equation for photon or neutrino transport (a high-dimensional parabolic PDE). A wide variety of algorithms for fluid dynamics are used, including finite-difference, finite-volume, and operator-splitting methods on orthogonal grids, as well as particle methods that are unique to astrophysics—for example, SPH. To improve resolution across a broad range of length scales, grid-based methods often rely on static and adaptive mesh refinement (AMR). The AMR methods greatly increase the complexity of the algorithm, reduce the scalability, and complicate effective load-balancing yet are absolutely essential for some problems.
- *Transport problems*. Required to calculate the effect of transport of energy and momentum by photons or neutrinos in a plasma. The mathematical model is a parabolic PDE in seven dimensions. Both grid-based (characteristic) and particle-based (Monte Carlo) methods are used. The high dimensionality of the problem makes first-principles calculations difficult, and so simplifying assumptions (for example, frequency-independent transport, or the diffusion approximation) are usually required.
- *Microphysics*. Necessary to incorporate nuclear reactions, chemistry, and ionization/recombination reactions into fluid and plasma simulations. The mathematical model is a set of coupled nonlinear, stiff ODEs (or algebraic equations if steady-state abundances are assumed) representing the reaction network. Implicit methods are generally required if the ODEs are solved. Implicit finite-difference methods for integrating realistic networks with dozens of constituent species are extremely costly.[4]

In its look at atmospheric sciences, that same report identified the following necessary computational advances:

[Advancing the state of atmospheric science research requires] the development of (1) scalable implementations of uniform-grid methods aimed at

---

[4] NRC, 2008. *The Potential Impact of High-End Capability Computing on Four Illustrative Fields of Sciences and Engineering*. The National Academies Press, Washington, D.C., p. 31.

the very highest performance and (2) a new generation of local refinement methods and codes for atmospheric, oceanic, and land modeling. . . . The propagation of uncertainty through a coupled model is particularly problematic, because nonlinear interactions can amplify the forced response of a system. In addition, it is often the case that we are interested in bounding the uncertainties in predictions of extreme, and hence rare, events, requiring a rather different set of statistical tools than those to study means and variances of large ensembles. New systematic theories about multiscale, multiphysics couplings are needed to quantify relationships better. This will be important as atmospheric modeling results are coupled with economic and impact models. Building a better understanding of coupling and the quantification of uncertainties through coupled systems is necessary groundwork for supporting the decisions that will be made based on modeling results.[5]

ILLUSTRATIVE DEMANDS FROM ENGINEERING

The 2008 National Academy of Engineering report *Grand Challenges for Engineering*[6] identified 14 major challenges. Following is a list of 11 of those challenges that depend on corresponding advances in the mathematical sciences, along with thoughts about the particular mathematical science research that will be needed.

- *Make solar energy economical.* This will require multiscale modeling of heterogeneous materials and better algorithms for modeling quantum-scale behaviors, and the mathematical sciences will contribute to both.
- *Provide energy from fusion.* This will require better methods for simulating multiscale, complex behavior, including turbulent flows, a topic challenging both mathematical scientists and domain scientists and engineers.
- *Develop carbon sequestration methods.* This will require better models of porous media and methods for modeling very large-scale heterogenous and multiphysics systems.
- *Advance health informatics.* Requires statistical research to enable more precise and tailored inferences from increasing amounts of data.
- *Engineer better medicines.* Requires tools for bioinformatics and simulation tools for modeling molecular interactions and cellular machinery.
- *Reverse-engineer the brain.* Requires tools for network analysis, models of cognition and learning, and signal analysis.

[5] Ibid. p. 58.
[6] From https://www.coursera.org/landing/hub.php.

- *Prevent nuclear terror.* Network analysis can contribute to this, as can cryptology, data mining, and other intelligence tools.
- *Secure cyberspace.* Requires advances in cryptography and theoretical computer science.
- *Enhance virtual reality.* Requires improved algorithms for scene rendering and simulation.
- *Advance personalized learning.* Requires advances in machine learning.
- *Engineer the tools of scientific discovery.* Requires advances to enable multiscale simulations, including improved algorithms, and improved methods of data analysis.

Engineering in general is dependent on the mathematical sciences, and that dependency is strengthening as engineers push toward ever greater precision. As just one illustration of the pervasiveness of the mathematical sciences in engineering, note the following examples of major improvements in manufacturing that were provided at the 2011 annual meeting of the National Academy of Engineering by Lawrence Burns, retired vice president for R&D and strategic planning at General Motors Corporation:

1. Simultaneous engineering,
2. Design for manufacturing,
3. Math-based design and engineering (computer-aided design),
4. Six-sigma quality,
5. Supply chain management,
6. The Toyota production system, and
7. Life-cycle analysis.[7]

It is striking that six of the seven items are inextricably linked with the mathematical sciences. While it is obvious that Item 3 depends on mathematical advances and Item 4 relies on statistical concepts and analyses, Items 1, 2, 5, and 7 are all dependent on advances in simulation capabilities that have enabled them to represent processes of ever-increasing complexity and fidelity. Research in the mathematical sciences is necessary in order to create those capabilities.

ILLUSTRATIVE DEMANDS FROM NETWORKING AND INFORMATION TECHNOLOGY

The 2010 report from the President's Council of Advisors on Science and Technology (PCAST) *Designing a Digital Future: Federally Funded*

---

[7] From http://www.nae.edu/Activities/Events/35831/51040/53248.aspx.

Research and Development in Networking and Information Technology[8] identified four major recommendations for "initiatives and investments" in networking and information technology (NIT) to "achieve America's priorities and advance key NIT research frontiers." Three of those initiatives are dependent on ongoing research in areas of the mathematical sciences. First, the mathematical sciences are fundamental to advances in simulation and modeling:

> The federal government should invest in a national, long-term, multi-agency, multi-faceted research initiative on NIT for energy and transportation. . . . Current research in the computer simulation of physical systems should be expanded to include the simulation and modeling of proposed energy-saving technologies, as well as advances in the basic techniques of simulation and modeling.[9]

Second, the mathematical sciences underpin cryptography and provide tools for systems analyses:

> The federal government should invest in a national, long-term, multi-agency research initiative on NIT that assures both the security and the robustness of cyber-infrastructure
>
> —to discover more effective ways to build trustworthy computing and communications systems,
> —to continue to develop new NIT defense mechanisms for today's infrastructure, and most importantly,
> —to develop fundamentally new approaches for the design of the underlying architecture of our cyber-infrastructure so that it can be made truly resilient to cyber-attack, natural disaster, and inadvertent failure.[10]

Third, the mathematical sciences are strongly entwined with R&D for privacy protection, large-scale data analysis, high-performance computing, and algorithms:

> The federal government must increase investment in those fundamental NIT research frontiers that will accelerate progress across a broad range of priorities . . . [including] research program on the fundamentals of privacy protection and protected disclosure of confidential data . . . fundamental

---

[8] PCAST, 2010, *Report to the President and Congress: Designing a Digital Future: Federally Funded Research and Development in Networking and Information Technology*, p. xi. Available at http://www.whitehouse.gov/sites/default/files/microsites/ostp/pcast-nitrd-report-2010.pdf.

[9] Ibid., p. xi.

[10] Ibid., p. xii.

research in data collection, storage, management, and automated large-scale analysis based on modeling and machine learning . . . [and] continued investment in important core areas such as high performance computing, scalable systems and networking, software creation and evolution, and algorithms.[11]

The report goes on to point out that five broad themes cut across its recommendations, including the need for capabilities to exploit ever-increasing amounts of data, to improve cybersecurity, and to better ensure privacy. As noted above, progress on these fronts will require strong mathematical sciences research.

More generally, the mathematical sciences play a critical role in any science or engineering field that involves network structures. In the early years of U.S. telecommunications, very few networks existed, and each was under the control of a single entity (such as AT&T). For more than 50 years in this environment, relatively simple mathematical models of call traffic were extremely useful for network management and planning. The world is completely different today: It is full of diverse and overlapping networks, including the Internet, the World Wide Web, wireless networks operated by multiple service providers, and social networks, as well as networks arising in scientific and engineering applications, such as networks that describe intra- and intercellular processes in biology. Modern technologies built around the Internet are consumers of mathematics: for example, many innovations at Google, such as search, learning, and trend discovery, are based on the mathematical sciences.

In settings ranging from the Internet to transportation networks to global financial markets, interactions happen in the context of a complex network.[12] The most striking feature of these networks is their size and global reach: They are built and operated by people and agents of diverse goals and interests. Much of today's technology depends on our ability to successfully build and maintain systems used by such diverse set of users, ensuring that participants cooperate despite their diverse goals and interests. Such large and decentralized networks provide amazing new opportunities for cooperation, but they also present large challenges. Despite the enormous amount of data collected by and about today's networks, fundamental mathematical science questions about the nature, structure, evolution, and security of networks remain that are of great interest for the government, for innovation-driven businesses, and for the general public.

---

[11] Ibid., p. xiii.

[12] The remainder of this section draws from pages 5, 6, and 10 of Institute for Pure and Applied Mathematics, 2012, *Report from the Workshop on Future Directions in Mathematics*. IPAM, Los Angeles.

For example, the sheer size of these networks makes it difficult to study them. The interconnectedness of financial networks makes financial interactions easy, but the financial crisis of 2008 provides a good example of the dangers of that interconnectivity. Understanding to what extent networks are susceptible to such cascading failures is an important area of study. From a very different domain, the network structure of the Web is what makes it most useful: Links provide the Web with a network structure and help us navigate it. But linking is also an endorsement of some sort. This network of implicit endorsements is what allows search companies such as Google to effectively find useful pages. Understanding how to harness such a large network of recommendations continues to be a challenge.

Game theory provides a mathematical framework that helps us understand the expected effects of interactions and to develop good design principles for building and operating such networks. In this framework we think of each participant as a player in a noncooperative game wherein each player selects a strategy, selfishly trying to optimize his or her own objective function. The outcome of the game for each participant depends, not only on his own strategy, but also on the strategies chosen by all other players. This emerging area is combining tools from many mathematical areas, including game theory, optimization, and theoretical computer science. The emergence of the Web and online social systems also gives graph theory an important new application domain.

ILLUSTRATIVE DEMANDS FROM BIOLOGY

The 2009 NRC report *A New Biology for the 21st Century*[13] had a great deal to say about emerging opportunities that will rely on advances in the mathematical sciences. Following are selected quotations:

- Fundamental understanding will require . . . computational modeling of their growth and development at the molecular and cellular levels. . . . The New Biology—integrating life science research with physical science, engineering, computational science, and mathematics—will enable the development of models of plant growth in cellular and molecular detail. Such predictive models, combined with a comprehensive approach to cataloguing and appreciating plant biodiversity and the evolutionary relationships among plants, will allow scientific plant breeding of a new type, in which genetic changes can be targeted in a way that will predictably result in novel crops and crops adapted to their conditions of growth. . . . New quantitative methods—the methods of the New Biology—are being developed that use next-generation DNA sequenc-

---

[13] National Research Council, 2009, *A New Biology for the 21st Century*. The National Academies Press, Washington, D.C.

ing to identify the differences in the genomes of parental varieties, and to identify which genes of the parents are associated with particular desired traits through quantitative trait mapping.[14]

- Fundamental advances in knowledge and a new generation of tools and technologies are needed to understand how ecosystems function, measure ecosystem services, allow restoration of damaged ecosystems, and minimize harmful impacts of human activities and climate change. What is needed is the New Biology, combining the knowledge base of ecology with those of organismal biology, evolutionary and comparative biology, climatology, hydrology, soil science, and environmental, civil, and systems engineering, through the unifying languages of mathematics, modeling, and computational science. This integration has the potential to generate breakthroughs in our ability to monitor ecosystem function, identify ecosystems at risk, and develop effective interventions to protect and restore ecosystem function.[15]

- Recent advances are enabling biomedical researchers to begin to study humans more comprehensively, as individuals whose health is determined by the interactions between these complex structural and metabolic networks. On the path from genotype to phenotype, each network is interlocked with many others through intricate interfaces, such as feedback loops. Study of the complex networks that monitor, report, and react to changes in human health is an area of biology that is poised for exponential development. . . . Computational and modeling approaches are beginning to allow analysis of these complex systems, with the ultimate goal of predicting how variations in individual components affect the function of the overall system. Many of the pieces are identified, and some circuits and interactions have been described, but true understanding is still well beyond reach. Combining fundamental knowledge with physical and computational analysis, modeling and engineering, in other words, the New Biology approach, is going to be the only way to bring understanding of these complex networks to a useful level of predictability.[16]

- Such complex events as how embryos develop or how cells of the immune system differentiate . . . must be viewed from a global yet detailed perspective because they are composed of a collection of molecular mechanisms that include junctions that interconnect vast networks of genes. It is essential to take a broader view and analyze entire gene regulatory networks, and the circuitry of events underlying complex biological systems. . . . Analysis of developing and differentiating systems at a network level will be critical for understanding complex events of how tissues and organs are assembled. . . . Similarly, networks of proteins interact at a biochemical level to form complex metabolic machines that produce distinct cellular products. Understanding these

---

[14] Ibid., pp. 22-23.
[15] Ibid., p. 25.
[16] Ibid., pp. 34-35.

and other complex networks from a holistic perspective offers the possibility of diagnosing human diseases that arise from subtle changes in network components.[17]

- Perhaps the most complex, fascinating, and least understood networks involve circuits of nerve cells that act in a coordinated fashion to produce learning, memory, movement, and cognition. . . . Understanding networks will require increasingly sophisticated, quantitative technologies to measure intermediates and output, which in turn may demand conceptual and technical advances in mathematical and computational approaches to the study of networks.[18]

Later, the report discusses some specific ways in which the mathematical sciences are enabling the New Biology:

[M]athematical underpinnings of the field . . . embrace probabilistic and combinatorial methods. Combinatorial algorithms are essential for solving the puzzles of genome assembly, sequence alignment, and phylogeny construction based on molecular data. Probabilistic models such as Hidden Markov models and Bayesian networks are now applied to gene finding and comparative genomics. Algorithms from statistics and machine learning are applied to genome-wide association studies and to problems of classification, clustering and feature selection arising in the analysis of large-scale gene expression data.[19]

As another illustration, the 2008 NRC report *The Potential Impact of High-End Capability Computing on Four Illustrative Fields of Science and Engineering*[20] identified essential challenges for evolutionary biology, a number of which are strongly mathematical. Following are selected quotations:

- Standard phylogenetic analysis comparing the possible evolutionary relationships between two species can be done using the method of maximum parsimony, which assumes that the simplest answer is the best one, or using a model-based approach. The former entails counting character change on alternative phylogenetic trees in order to find the tree that minimizes the number of character transformations. The latter incorporates specific models of character change and uses a minimization criterion to choose among the sampled trees, which often involves finding the tree with the highest likelihood. Counting, or optimizing, change on a tree, whether in a parsimony or model-based framework, is a computationally efficient problem. But sampling all possible trees to

---

[17] Ibid., p. 35.

[18] Ibid.

[19] Ibid., p. 62.

[20] NRC, 2008, *The Potential Impact of High-End Capability Computing on Four Illustrative Fields of Science and Engineering*. The National Academies Press, Washington, D.C.

find the optimal solution scales precipitously with the number of taxa (or sequences) being analyzed. . . . Thus, it has long been appreciated that finding an exact solution to a phylogenetic problem of even moderate size is NP complete. . . . Accordingly, numerous algorithms have been introduced to search heuristically across tree space and are widely employed by biological investigators using platforms that range from desktop workstations to supercomputers. These algorithms include methods [such as] simulated annealing, genetic (evolutionary) algorithmic searches, and Bayesian Markov chain Monte Carlo (MCMC) approaches.[21]

- The accumulation of vast amounts of DNA sequence data and our expanding understanding of molecular evolution have led to the development of increasingly complex models of molecular evolutionary change. As a consequence, the enlarged parameter space required by these molecular models has increased the computational challenges confronting phylogeneticists, particularly in the case of data sets that combine numerous genes, each with their own molecular dynamics. . . . Phylogeneticists are more and more concerned about having statistically sound measures of estimating branch support. In model-based approaches, in particular, such procedures are computationally intensive, and the model structure scales significantly with the size of the number of taxa and the heterogeneity of the data. In addition, more attention is being paid to statistical models of molecular evolution.[22]

- A serious major computational challenge is to generate qualitative and quantitative models of development as a necessary prelude to applying sophisticated evolutionary models to understand how developmental processes evolve. Developmental biologists are just beginning to create the algorithms they need for such analyses, based on relatively simple reaction-rate equations, but progress is rapid. . . . Another important breakthrough in the field is the analysis of gene regulatory networks [which] describe the pathways and interactions that guide development.[23]

- Modern sequencing technologies routinely yield relatively short fragments of a genomic sequence, from 25 to 1,000 [base pairs]. Whole genomes range in size from the typical microbial sequence, which has millions of base pairs, to plant and animal sequences, which often consist of billions of base pairs. Additionally, 'metagenomic' sequencing from environmental samples often mixes fragments from dozens to hundreds of different species and/or ecotypes. The challenge is to take these short subsequences and assemble them to reconstruct the genomes of species and/or ecosystems. . . . While the fragment assembly problem is NP-complete, heuristic algorithms have produced high-quality reconstructions of hundreds of genomes. The recent trend is toward

---

[21] Ibid., p. 65.
[22] Ibid.
[23] Ibid., p. 72.

methods of sequencing that can inexpensively generate large numbers (hundreds of millions) of ultrashort sequences (25-50 bp). Technical and algorithmic challenges include the following:

—Parallelization of all-against-all fragment alignment computations.

—Development of methods to traverse the resulting graphs of fragment alignments to maximize some feature of the assembly path.

—Heuristic pruning of the fragment alignment graph to eliminate experimentally inconsistent subpaths.

—Signal processing of raw sequencing data to produce higher quality fragment sequences and better characterization of their error profiles.

—Development of new representations of the sequence-assembly problem—for example, string graphs that represent data and assembly in terms of words within the dataset.

—Alignment of error-prone resequencing data from a population of individuals against a reference genome to identify and characterize individual variations in the face of noisy data.

—Demonstration that the new methodologies are feasible, by producing and analyzing suites of simulated data sets.[24]

- Once we have a reconstructed genomic or metagenomic sequence, a further challenge is to identify and characterize its functional elements: protein-coding genes; noncoding genes, including a variety of small RNAs; and regulatory elements that control gene expression, splicing, and chromatin structure. Algorithms to identify these functional regions use both statistical signals intrinsic to the sequence that are characteristic of a particular type of functional region and comparative analyses of closely and/or distantly related sequences. Signal-detection methods have focused on hidden Markov models and variations on them. Secondary structure calculations take advantage of stochastic, context-free grammars to represent long-range structural correlations.[25]

- Comparative methods require the development of efficient alignment methods and sophisticated statistical models for sequence evolution that are often intended to quantitatively model the likelihood of a detected alignment given a specific model of evolution. While earlier models treated each position independently, as large data sets became available the trend is now to incorporate correlations between sites. To compare dozens of related sequences, phylogenetic methods must be integrated with signal detection.[26]

ILLUSTRATIVE DEMANDS FROM MEDICINE

There is increasing awareness in medicine of the benefits from collaboration with mathematical scientists, including in areas where past focus has

---

[24] Ibid., pp. 77-78.

[25] Ibid.

[26] Ibid.

tended to be clinical and to have little reference to mathematics. In areas ranging from medical imaging, drug discovery, discovery of genes linked to hereditary diseases, personalized medicine, validating treatments, cost-benefit analysis, and robotic surgery, the mathematical sciences are ever-present. The role of the mathematical sciences in medicine has become so important that one can provide only a few examples.

One example is a program of the Office of Physcal Sciences Oncology,[27] funded by the National Cancer Institute, to support innovative approaches to understanding and controlling cancer, which is explicitly seeking teams that may include mathematical scientists. Some of the associated mathematical science questions arise in creation of sophisticated models of cancer, optimization of cancer chemotherapy regimes, and rapid incorporation of data into models and treatment. (As shown in Appendix C, the National Institutes of Health overall allocates some $90 million per year for mathematical sciences research.) As another example, looking just at the February 2011 issue of *Physical Biology*, one sees articles whose titles contain the following mathematics-related phrases: "stochastic dynamics," "micromechanics," "an evolutionary game theoretical view," and "an Ising model of cancer and beyond." In the same vein, a research highlight published in *Cancer Discovery* in July 2011 describes development of "a statistical approach that maps out the order in which . . . [cancer-related] abnormalities arise." It is increasingly common to find such cross-fertilization.

RESTORING REPRODUCIBILITY IN SCIENCE

A recent article in the *Wall Street Journal* opens with the following story:

> Two years ago, a group of Boston researchers published a study describing how they had destroyed cancer tumors by targeting a protein called STK33. Scientists at biotechnology firm Amgen Inc. quickly pounced on the idea and assigned two dozen researchers to try to repeat the experiment with a goal of turning the findings into a drug. It proved to be a waste of time and money. After six months of intensive lab work, Amgen found it couldn't replicate the results and scrapped the project. 'I was disappointed but not surprised,' says Glenn Begley, vice president of research at Amgen of Thousand Oaks, Calif. 'More often than not, we are unable to reproduce findings' published by researchers in journals.[28]

---

[27] The Office of Physical Sciences in Oncology; http://physics.cancer.gov/about/summary.asp.

[28] Gautam Naik, Scientists' elusive goal: Reproducing study results. *Wall Street Journal*, December 2, 2011.

A related paper[29] reports on a study done by Bayer Healthcare that examined 67 published research articles for which Bayer had attempted to replicate the findings in-house. Fewer than one-quarter were viewed as having been essentially replicated, while more than two-thirds exhibited major inconsistencies, in most cases leading Bayer to terminate the projects.

One reason for this lack of reproducibility is undoubtedly the pressure to publish and find positive results. Under this pressure, contradictory evidence can be swept under the carpet with rationalizations about why such evidence may safely be ignored. And, of course, some experiments will simply have been done wrong, having undetected sources of bias (e.g., contaminated materials).

Another reason for the lack of reproducibility is unsound statistical analysis. Causes range from scientifically improper selection of data (throwing out data that do not support the claim) to a weak understanding of even the simplest statistical methods. As an example of the latter, Nieuwenhuis et al.[30] reviewed 513 behavioral, systems, and cognitive neuroscience articles in five top-ranking journals, looking for articles that specifically compared two experimental effects to see if they differed significantly. Out of the 157 articles that focused on such a comparison, they found that 78 used the correct statistical procedure of testing whether the difference of the two effects is significantly different from 0; 79 used a blatantly incorrect procedure. The incorrect procedure was to separately test each experimental effect against a null hypothesis of no effect and, if one effect was significantly different from zero but the other was not, to declare that there was a significant difference between the effects. Of course, this can happen even when the two effects are essentially identical (e.g., if the p-value for one is 0.06 and the p-value for the other is 0.05).

Ensuring sound statistical analysis has long been a problem for science, and this problem is only getting worse owing to the increasingly massive data available today which allows increasingly thorough exploration of the data for "artifacts." These artifacts can be exciting new science or they can simply arise from noise in the data. The worry these days is that most often it is the latter.

In a recent talk about the drug discovery process (which the committee chooses not to cite in order to avoid embarrassing the presenter), the following numbers were given in illustration:

---

[29] Florian Prinz, Thomas Schlange, and Khusru Asadullah, 2011, Believe it or not: How much can we rely on published data on potential drug targets? *Nature Reviews Drug Discovery* 10:712.

[30] Sander Nieuwenhuis, Birte U Forstmann, and Eric-Jan Wagenmakers, 2011, Erroneous analyses of interactions in neuroscience: a problem of significance. *Nature Neuroscience* 14: 1105-1107.

- Ten thousand relevant compounds were screened for biological activity.
- Five hundred passed the initial screen and were subjected to in vitro experiments.
- Twenty-five passed this screening and were studied in Phase I animal trials.
- One passed this screening and was studied in an expensive Phase II human trial.

These numbers are completely compatible with the presence of nothing but noise, assuming the screening was done based on statistical significance at the 0.05 level, as is common. (Even if none of the 10,000 compounds has any effect, roughly 5 percent of the compounds would appear to have an effect at the first screening; roughly 5 percent of the 500 screened compounds would appear to have an effect at the second screening, etc.)

This problem, often called the "multiplicity" or "multiple testing" problem, is of central importance not only in drug development, but also in microarray and other bioinformatic analyses, syndromic surveillance, high-energy physics, high-throughput screening, subgroup analysis, and indeed any area of science facing an inundation of data (which most of them are). The section of Chapter 2 on high-dimensional data indicates the major advances happening in statistics and mathematics to meet the challenge of multiplicity and to help restore the reproducibility of science.

Appendix E

Illustrative Programs Aimed at Increasing Participation in the Mathematical Sciences by Women and Underrepresented Minorities

PROGRAMS AIMED AT K-12 FEMALES

All Girls/All Math
(http://www.math.unl.edu/programs/agam/)

The University of Nebraska-Lincoln offers a weeklong summer camp for high school girls who are interested in exploring mathematical topics that are outside the scope of the typical high school curriculum. Participants take a one-week course and attend lectures on other exciting topics in the mathematical sciences. The camp is structured so that the participants are taught by female mathematics professors and chaperoned by female mathematics graduate students. There is also a keynote presentation by a prominent female mathematician.

Sonia Kovalevsky High School Mathematics Days
(http://sites.google.com/site/awmmath/programs/kovalevsky-days)

The Association for Women in Mathematics (AWM) supports Sonia Kovalevsky High School and Middle School Mathematics Days by providing grants up to $3,000 to colleges and universities nationwide. This program, which is currently supported by a grant from the National Science Foundation (NSF), consists of a program of workshops, talks, and problem-solving competitions for female high school or middle school students and their teachers. The purpose of Sonia Kovalevsky Days is to "encourage young women to continue their study of mathematics, to assist them with the sometimes difficult transitions between middle school and high school mathematics and between high school and college mathematics, to assist the teachers of women mathematics students, and to encourage colleges and universities to develop more extensive cooperation with middle schools and high schools in their area."

Tensor Women and Mathematics Grants
(http://www.maa.org/wam/tensor.html)

The Mathematics Association of America awards grants of up to $6,000 to college or university and secondary mathematics faculty for projects designed to encourage college and university women or high school and middle school girls to study mathematics. This program, funded by the Tensor Foundation, supports activities such as the following:

- Organizing a club for women interested in mathematics or mathematics and science;
- Creating a network of women professional mentors who will direct mathematics projects for girls;
- Holding a conference for counselors to prepare them to encourage women and girls to continue to study mathematics;
- Conducting a summer mathematics program for high school women;
- Bringing high school women onto a college campus for a "math day" with follow-up;
- Structuring a program for high school and/or college women to mentor younger female mathematics students with math projects or math clubs;
- Forming partnerships with industry to acquaint women students with real-life applications of mathematics; and
- Providing funds toward release time to allow a faculty member to prepare a course on women and mathematics, provided the host institution agrees to offer such a course.

PROGRAMS AIMED AT UNDERGRADUATE
AND GRADUATE FEMALE STUDENTS

Alice T. Schafer Prize
(http://sites.google.com/site/awmmath/programs/schafer-prize)

The Alice T. Schafer Mathematics Prize of the Association for Women in Mathematics (AWM) is awarded to an undergraduate woman residing in the United States for excellence in mathematics.

AWM Student Chapters
(http://sites.google.com/site/awmmath/programs/student-chapters)

AWM student chapters hold regular meetings and events open to all undergraduate and graduate students, regardless of major or gender. These meetings and activities allow students to be exposed to the world of professional mathematics, to obtain information about the varied career options in mathematics, to network with professional mathematicians, and to develop leadership skills. Activities for student chapters include sponsoring a lecture series by either students or local mathematicians, site visits to major employers of mathematicians, outreach through activities such as tutoring, social gatherings such as picnics or banquets, mentoring programs for youth, and special events such as career days.

Workshops for Women Graduate Students and Postdoctoral Mathematicians
(http://sites.google.com/site/awmmath/programs/workshops)

AWM holds a series of workshops for women graduate students and recent Ph.D.s in conjunction with major mathematics meetings. These workshops consist of a graduate student poster session, presentations by recent Ph.D.s, mentoring events, and career symposia. Workshop participants have the opportunity to meet with other women mathematicians at all stages of their careers.

Travel Grants for Women Researchers
(http://sites.google.com/site/awmmath/programs/travel-grants)

AWM administers travel grants for women to support attendance at research conferences and longer-term visits with a mentor. The purpose of the travel grants is to enhance the research activities of women mathematicians and increase their visibility in various research venues.

Mentoring Travel Grants for Women
(http://sites.google.com/site/awmmath/programs/
travel-grants/mathematics-mentoring-travel-grants)

AWM provides mathematics mentoring travel grants to help junior women to develop long-term working and mentoring relationships with senior mathematicians. This relationship should help the junior mathematician to establish her research program and eventually receive tenure. Each grant funds travel, accommodations, and other required expenses for an untenured woman mathematician to travel to an institute or a department to do research with a specified individual for one month. The research areas of both the applicant and the mentor must be in a field that is supported by NSF/Division of Mathematical Sciences.

PROGRAMS AIMED AT K-12 UNDERREPRESENTED MINORITIES

Joaquin Bustoz Math-Science Honors Program
(http://mshp.asu.edu/)

Arizona State University (ASU) runs an intense academic program that provides an opportunity for underrepresented minority students to begin university mathematics and science studies before graduating from high school. All expenses are paid by ASU and participants live on the ASU Tempe campus while enrolled in a university-level mathematics course for college credit.

Louisiana Preparatory (LaPREP) Program
(http://www.lsus.edu/offices-and-services/
community-outreach/laprep-program)

LaPREP is a two-summer enrichment program that identifies, encourages, and instructs competent middle and early high school students, preparing them to complete a college degree program in math, science, or engineering. Participants attend 7 weeks of intellectually demanding classes and seminars, with emphasis on abstract reasoning, problem solving, and technical writing skills interspersed with field trips to local industries. Some of the mathematical topics studied over two summer sessions include logic, algebraic structures, and probability and statistics.

Mathematics, Engineering, Science Achievement (MESA) Program
(http://mesa.ucop.edu/)

The MESA program serves African-American, American-Indian, Mexican-American, and other Latino-Americans in California who have been historically underrepresented in mathematics-based fields. MESA, which is funded by the University of California at Berkeley and the California State University (CSU) Chancellor's Office, offers activities such as field trips and enrichment classes for junior high and high school students. Each year there is a MESA day, when a science olympiad is held.

Mathematics Intensive Summer Session (MISS)
(http://www.fullerton.edu/sa/miss/)

MISS, a 4-week commuter program held during the summer at California State University at Fullerton, is designed to help female minority high school students be successful in their college preparatory mathematics classes. Participants learn about topics in Algebra II as preparation for taking Algebra II or integrated Math III when they return to school in the fall. All expenses are paid by the California State University at Fullerton.

Tensor-SUMMA Grants: Strengthening Underrepresented Minority Mathematics Achievement (http://www.fullerton.edu/sa/miss/)

The Mathematical Association of America awards up to $6,000 to college and university mathematical science faculty and departments for programs designed to encourage the pursuit and enjoyment of mathematics among students in middle school, high school, or beginning in college from groups traditionally underrepresented in the field of mathematics. This program, funded by the Tensor Foundation, supports activities such as preparation for competitions within the mathematical sciences, math circles, student group and individual research experiences, summer mathematics camp, and math club activities.

The Texas Prefreshman Engineering Program (http://www.prep-usa.org/portal/texprep/)

The Texas Prefreshman Engineering Program (TexPrep) identifies achieving middle and high school students with an interest in engineering, science, technology, and other mathematics-related areas and strengthens their potential for careers in these fields. TexPREP is a collaborative effort of colleges and universities throughout Texas that encourages students to begin preparing early for scientific and engineering career paths in school. Women and members of minority groups traditionally underrepresented in these areas continue to be target groups.

PROGRAMS AND ORGANIZATIONS AIMED AT UNDERGRADUATE AND GRADUATE STUDENTS FROM UNDERREPRESENTED MINORITIES

California State University Alliance for Minority Participation Project (http://students.ucsd.edu/academics/research/ undergraduate-research/opportunities/camp.html)

Many California state universities and partnering community colleges have a CSU-AMP program aimed at increasing the number of bachelor's degree recipients in science, engineering, and mathematics among historically underrepresented groups in these fields. CSU-AMP involves students in science, engineering, and mathematics enrichment activities throughout their entire undergraduate careers. While programs vary from campus to campus, students often receive two summers of intensive work in mathematics, academic-year workshops that support their mathematics and science courses, and research internships. This program is sponsored by the NSF.

California Alliance for Minority Participation

Many University of California campuses have California Alliance for Minority Participation (CAMP) programs that offer support and advancement opportunities to underrepresented students seeking bachelor's degrees in chemistry, engineering, mathematics, physics, or other sciences. CAMP participants can take advantage of events and services designed to meet the needs of a culturally and intellectually diverse group of students. Some CAMP activities include workshops, research projects, meetings with faculty, and scholarships. Participants must be a member of an underserved minority (African-American, Chicano, Latino, Native American, or Pacific Islander) and majoring in chemistry, engineering, mathematics, physics, or other sciences.

RESOURCES FOR FEMALE FACULTY MEMBERS

- Ruth I. Michler Memorial Prize (http://sites.google.com/site/awmmath/programs/michler-prize),
- Travel grants for women researchers (http://sites.google.com/site/awmmath/programs/travel-grants), and
- Mentoring travel grants for women (http://sites.google.com/site/awmmath/programs/travel-grants/mathematics-mentoring-travel-grants).

OTHER RESOURCES TO ENCOURAGE WOMEN IN THE MATHEMATICAL SCIENCES

- Organizing meetings in cooperation with AWM (http://sites.google.com/site/awmmath/in-cooperation-with),
- Humphreys Award (http://sites.google.com/site/awmmath/programs/humphreys-award),
- Louise Hay Award (http://sites.google.com/site/awmmath/programs/hay-award),
- Noether Lectures (http://sites.google.com/site/awmmath/programs/noether-lectures),

- Falconer Lectures (http://sites.google.com/site/awmmath/
  programs/falconer-lectures),
- Kovalevsky Lectures (http://sites.google.com/site/awmmath/
  programs/kovalevsky-lectures),
- Teacher partnership (http://sites.google.com/site/awmmath/
  programs/teacher-partnership), and
- Mentor network (http://sites.google.com/site/awmmath/programs/
  mentor-network).

Appendix F

Biographical Sketches of Committee Members and Staff

Thomas E. Everhart (Chair) is president emeritus of the California Institute of Technology and professor emeritus of electrical engineering and applied physics. He received a Ph.D. in engineering from Cambridge University, England, in 1958. Dr. Everhart joined the University of California at Berkeley in 1958, where he served in the Department of Electrical Engineering and Computer Science for more than 20 years. After serving as dean of engineering at Cornell University (1979-1984) and chancellor of the University of Illinois at Champaign-Urbana (1984-1987), he accepted the presidency of the California Institute of Technology (Caltech) in 1987. He holds a guest appointment at the University of California at Santa Barbara as a Distinguished Visiting Professor and Senior Advisor to the Chancellor. Dr. Everhart's honors and awards include the Institute of Electrical and Electronics Engineering (IEEE) Centennial Medal; the 1989 Benjamin Garver Lamme Award from the American Society for Engineering Education; the Clark Kerr Award from the University of California, Berkeley, in 1992; the Founder's Award in 1995 from the Energy and Resources Group at Berkeley; and the IEEE Founders Medal and Okawa Prize in 2002. A member of the U.S. National Academy of Engineering (NAE) and a Foreign Member of the Royal Academy of Engineering, he currently serves on the board of trustees of Caltech and has served on the board of overseers of Harvard University. He is currently a director of the W.M. Keck Foundation and the Kavli Foundation. He has consulted for industry at various times and has served on the boards of directors of General Motors, Hewlett-Packard, Hughes, Raytheon, and Saint-Gobain, among others.

Mark L. Green (**Vice-Chair**) is a professor in the Department of Mathematics at the University of California, Los Angeles (UCLA). He received a B.S. from the Massachusetts Institute of Technology (MIT) and his M.A. and Ph.D. from Princeton University. After teaching at the University of California, Berkeley, and MIT, he came to UCLA as an assistant professor in 1975. He was a founding codirector of the National Science Foundation NSF-funded Institute for Pure and Applied Mathematics. Dr. Green's research has taken him into different areas of mathematics: several complex variables, differential geometry, commutative algebra, Hodge theory, and algebraic geometry. He received an Alfred P. Sloan fellowship, was an invited speaker at the International Congress of Mathematicians in Berlin in 1998, and was recently elected as a fellow of the American Academy of Arts and Sciences, the American Association for the Advancement of Science, and the American Mathematical Society.

Tanya S. Beder is CEO and chairman of SBCC Group, a financial and risk advisory firm that she founded in 1987. She also serves as a director of American Century mutual fund complex in Mountain View, California, where she chairs the Risk Committee, and as a director of CYS Investments, a specialty finance company traded on the New York Stock Exchange. From 1994 through 2005, Ms. Beder held two senior positions in the asset management industry, first as managing director of Caxton Associates LLC, a $10 billion asset management firm, then as CEO of Tribeca Global Management LLC, a $3 billion dollar multistrategy fund. At SBCC Group, Ms. Beder heads the global strategy, crisis and risk management, derivatives, workout, and fund launch practices. Ms. Beder is a member of the board of directors of the International Association of Financial Engineers, where she co-chairs its Investor Risk Committee. From 1998 through 2003 she was chairman of that association. *Euromoney* named Ms. Beder one of the top 50 women in finance around the world, and *The Hedgefund Journal* named her one of the 50 leading women in hedge funds. While CEO of Tribeca, *Absolute Return* awarded her the prestigious Institutional Investment Manager of the Year Award. Ms. Beder is an author of the book *Financial Engineering, The Evolution of a Profession*, published in 2011, which discusses the uses and misuses of derivatives and complex instruments in global capital markets, and has written numerous articles in the financial arena. At Stanford University she teaches the course Strategy and Policy Issues in Financial Engineering. Previously Ms. Beder taught courses at Yale University's School of Management, Columbia University's Graduate School of Business and Financial Engineering, and the New York Institute of Finance. Ms. Beder also serves on advisory boards at Columbia University and New York University's Courant Institute and is an appointed Fellow of the International Center for Finance at Yale. She holds an M.B.A.

from Harvard University and a B.A. in mathematics and philosophy from Yale University. She was a member of the National Science Foundation's (NSF's) "Odom committee" in the late-1990s, which performed the last introspective study of the mathematical sciences.

James O. Berger is arts and sciences professor in the Department of Statistical Science at Duke University. He received a Ph.D. in mathematics from Cornell University in 1974. Dr. Berger was a faculty member in the Department of Statistics at Purdue University until 1997, at which time he moved to Duke. From 2002 until 2009 he directed the NSF-supported Statistical and Applied Mathematical Sciences Institute (SAMSI). Dr. Berger was president of the Institute of Mathematical Statistics in 1995 and 1996, chair of the Section on Bayesian Statistical Science of the American Statistical Association in 1995, and president of the International Society for Bayesian Analysis during 2004. Among his awards and honors are Guggenheim and Sloan Fellowships, the President's Award from the Committee of Presidents of Statistical Societies in 1985, the Sigma Xi Research Award at Purdue University for contribution of the year to science in 1993, the Fisher Lectureship in 2001, election as foreign member of the Spanish Real Academia de Ciencias in 2002, election to the U.S. National Academy of Sciences (NAS) in 2003, an honorary D.Sc. from Purdue University in 2004, and the Wald Lectureship in 2007. Professor Berger currently chairs the NSF's Advisory Committee for Mathematics and Physical Sciences. His research has primarily been in Bayesian statistics, foundations of statistics, statistical decision theory, simulation, model selection, and various interdisciplinary areas of science and industry, especially astronomy and the interface between computer modeling and statistics. He has supervised 31 Ph.D. dissertations, published over 160 articles, and written or edited 14 books or special volumes.

Luis A. Caffarelli is a professor of mathematics at the Institute for Computational Engineering and Sciences and holds the Sid W. Richardson Foundation Regents Chair in Mathematics at the University of Texas at Austin. He obtained and M.Sc. (1969) and a Ph.D. (1972) at the University of Buenos Aires. He also has been a professor at the University of Minnesota, the University of Chicago, and the Courant Institute of Mathematical Sciences at New York University (NYU). From 1986 to 1996 he was a professor at the Institute for Advanced Study in Princeton. In 1991, he was elected to the NAS. He received the Bôcher Prize in 1984. In 2005, he received the prestigious Rolf Schock Prize in Mathematics of the Royal Swedish Academy of Sciences. He recently received the Leroy P. Steele Prize for Lifetime Achievement in Mathematics. Professor Caffarelli is a member of the American Mathematical Society, the Union Matematica Argentina, the

American Academy of Arts and Sciences, and the Pontifical Academy of Sciences. The focus of Professor Caffarelli's research has been in the area of elliptic nonlinear partial differential equations and their applications. His research has reached from theoretical questions about the regularity of solutions to fully nonlinear elliptic equations to partial regularity properties of Navier-Stokes equations. Some of his most significant contributions are the regularity of free boundary problems and solutions to nonlinear elliptic partial differential equations, optimal transportation theory, and results in the theory of homogenization.

Emmanuel J. Candes is a professor of statistics and mathematics at Stanford University. He has carried out research into compressive sensing, mathematical signal processing, computational harmonic analysis, multi-scale analysis, scientific computing, statistical estimation and detection, high-dimensional statistics, theoretical computer science, mathematical optimization, and information theory. He received his Diplôme from the Ecole Polytechnique and his Ph.D. in statistics from Stanford University in 1998.

Phillip Colella is senior mathematician and goup leader of the Applied Numerical Algorithms Group at the E.O. Lawrence Berkeley National Laboratory. He is a leader in the development of mathematical methods and computer science tools for science and engineering. His work has resulted in software tools applicable in a wide variety of problems in fluid dynamics, shock wave theory, and astrophysics. Dr. Colella received an A.B. and a Ph.D. from the University of California at Berkeley. He was elected to the NAS in 2004.

David Eisenbud was director of the Mathematical Sciences Research Institute at the University of California at Berkeley, from 1997 until 2007, and he continues to serve on the faculty of Berkeley as professor of mathematics. In 2009 he also became director of mathematics and the physical sciences at the Simons Foundation. Dr. Eisenbud received his Ph.D. in mathematics in 1970 at the University of Chicago. He was on the faculty at Brandeis University for 27 years before coming to Berkeley and has also been a visiting professor at Harvard, Bonn, and Paris. His mathematical interests range widely over commutative and noncommutative algebra, algebraic geometry, topology, and computer methods. He was president of the American Mathematical Society (AMS) in 2004 and 2005 and is a director of Math for America, a foundation devoted to improving mathematics teaching. In 2006, Dr. Eisenbud was elected a fellow of the American Academy of Arts and Sciences. He currently serves on the editorial boards of the *Journal of Algebra and Number Theory*, the *Bulletin de la Société Mathématique de*

France, *Computing in Science & Engineering,* and Springer-Verlag's book series Algorithms and Computation in Mathematics.

Peter Wilcox Jones is the James E. English Professor of Mathematics and Applied Mathematics at Yale University. He received his doctorate from UCLA in pure mathematics in 1978 and began his lifelong international collaborations during his graduate studies, when he relocated to Paris during his advisor's year-long sabbatical to the University of Paris at Orsay. Dr. Jones began his academic career at the University of Chicago in 1978, where he served for 2 years as assistant director of the Institut Mittag-Leffler, a research branch of the Royal Swedish Academy of Sciences. He received the Salem Prize in 1981, an award given annually to a young mathematician who has done outstanding work in the theory of Fourier series. Dr. Jones joined the Department of Mathematics at Yale in 1985, where he currently works with a large group that focuses on the value of math in biology and medicine for creating models. Since its inception in 1999, he has served as the chair of the Science Advisory Board at IPAM, a mathematics research institute at UCLA created and funded by the NSF. Dr. Jones is a foreign member of the Swedish Academy of Science and a member of the American Academy of Arts and Sciences and the NAS.

Ju-Lee Kim is an associate professor of mathematics at MIT. She received a B.S. from the Korean Advanced Institute in Science & Technology in 1991 and a Ph.D. from Yale University in 1997. She had postdoctoral appointments at the École Normale Supérieure and the Institute for Advanced Study before joining the University of Michigan as assistant professor in 1998. In 2002, she moved to the University of Illinois at Chicago before joining MIT. Dr. Kim's research interests include representation theory, harmonic analysis of p-adic groups, Lie theory, and automorphic forms.

Yann LeCun has been a professor of computer science at the Courant Institute of Mathematical Sciences at NYU since 2003 and was named Silver Professor in 2008. Dr. LeCun received a Ph.D. in computer science from the Université Pierre et Marie Curie, Paris in 1987. He joined the Adaptive Systems Research Department at AT&T Bell Laboratories in Holmdel, New Jersey, in 1988, where he later became head of the Image Processing Research Department within the Speech and Image Processing Research Lab at AT&T Labs-Research. In 2002, he became a fellow of the NEC Research Institute (now NEC Labs America) in Princeton, New Jersey. Dr. LeCun's research focuses on machine learning, computer vision, pattern recognition, neural networks, handwriting recognition, image compression, document understanding, image processing, VLSI design, and information theory. His handwriting recognition technology is used by several banks around the world, and his image compression technology is used by hundreds of Web sites and publishers and millions of users to access scanned documents on the Web.

Jun Liu is a professor of statistics at Harvard University and of biostatistics in the Harvard School of Public Health. His research deals with statistical imputation, Gibbs sampling, graphical models, genetics, image reconstructions, and other methods of biostatistics and bioinformatics. He holds a B.S. in mathematics from Peking University (1985) and a Ph.D. in statistics from the University of Chicago (1991). Dr. Liu began his career at Harvard in 1991, was at Stanford in 1994-2000, and returned to Harvard in 2000. His honors include selection as a Medallion lecturer of the Institute for Mathematical Statistics (IMS) in 2002; receipt of the 2002 COPSS Presidents' Award, given annually by five leading statistical societies to a young individual for outstanding contributions to the profession of statistics; election as an IMS fellow in 2004; and selection as Bernoulli lecturer by the Bernoulli Society, 2004. He is the author of *Monte Carlo Strategies in Scientific Computing* (2001), has overseen 18 Ph.D. students, and has contributed to 18 software modules for computational biology.

Juan Maldacena is a theoretical physicist at the Institute for Advanced Study in Princeton, New Jersey. Among his many discoveries, the most famous is the AdS/CFT correspondence, the conjecture about the equivalence of string theory on Anti de Sitter (AdS) space and a conformal field theory defined on the boundary of the AdS space. Dr. Maldacena obtained his "licenciatura" (a 6-year degree) in 1991 from the Instituto Balseiro from the Universidad Nacional de Cuyo, Bariloche, Argentina. He then obtained his Ph.D. at Princeton University in 1996 and went on to a postdoctoral position at Rutgers University. In 1997, he joined Harvard University as associate professor, being promoted to professor of physics in 1999. Since 2001 he has been a professor at the Institute for Advanced Study. He has the Edward A. Bouchet Award of the American Physical Society (2004), the Xanthopoulos International Award for Research in Gravitational Physics (2001), the Sackler Prize in Physics, a MacArthur fellowship in 1999, and the Dirac medal in 2008.

John W. Morgan is director of the Simons Center for Geometry and Physics at SUNY-Stony Brook. He received his B.A. and Ph.D. in mathematics from Rice University in 1968 and 1969, respectively. He was an instructor at Princeton University from 1969 to 1972 and an assistant professor at MIT from 1972 to 1974. He has been on the faculty at Columbia University since 1974. In July 2009, Dr. Morgan moved to Stony Brook University to become the first director of the Simons Center for Geometry and Physics. He has been a visiting professor at Harvard University, Stanford University, the Université de Paris, MSRI, the Institute for Advanced Study, and the Institut des Hautes Etudes Scientifiques. A member of the NAS, he is an editor of the *Journal of the American Mathematical Society* and *Geometry and Topology*.

Yuval Peres obtained his Ph.D. in 1990 from the Hebrew University in Jerusalem, working under Hillel Furstenberg. In 1993, Dr. Peres joined the faculty of the statistics department at the University of California, Berkeley, where he served as a professor in the mathematics and statistics departments until moving recently to Microsoft Research to manage the Theory Group. Peres's research encompasses a broad range of topics in theoretical probability. His research could be characterized as probability on infinite discrete structures where geometry plays a role. This includes, for instance, the study of random percolation on infinite Cayley graphs, where (in contrast to the usual d-dimensional lattice setting) one has the possibility of coexistence of infinitely many infinite components. Peres's work illustrates and delineates active and exciting areas where probability meets other areas of pure mathematics.

Eva Tardos is the Jacob Gould Schurman Professor of Computer Science at Cornell University and was department chair 2006-2010. She received her B.A. and Ph.D. from Eötvös University in Budapest. She had a Humboldt fellowship at the University of Bonn, postdoctoral fellowships at MSRI and from the Hungarian Academy of Sciences at Eötvös University, and was a visiting professor at the Department of Mathematics at MIT in 1987-1989 before joining the faculty at Cornell. Dr. Tardos won the Fulkerson Prize, awarded jointly by the Mathematical Programming Society (MPS) and the AMS, and the Dantzig prize, awarded jointly by MPS and the Society for Industrial and Applied Mathematics (SIAM). She was awarded an Alfred P. Sloan Research Fellowship (1991-1993), an NSF Presidential Young Investigator Award (1991-1996), the David and Lucille Packard Foundation Fellowship in Science and Engineering (1990-1995), and a Guggenheim fellowship (1999-2000). She is a fellow of the Association for Computing Machinery (ACM), INFORMS, and SIAM, and is a member of the American Academy of Arts and Sciences and the NAE. Dr. Tardos's research interest is algorithms and algorithmic game theory, the subarea of computer science theory that involves designing systems and algorithms for selfish users. Her research focuses on algorithms and games on networks. She is most known for her work on network-flow algorithms, approximation algorithms, and quantifying the efficiency of selfish routing.

Margaret H. Wright is Silver Professor of Computer Science at the Courant Institute of Mathematical Sciences, NYU. She received an M.S. and a Ph.D.

in computer science and a B.S. in mathematics, all from Stanford University. Before joining NYU in 2001, she was a Distinguished Member of the Technical Staff and Bell Labs fellow at Bell Laboratories, Lucent Technologies. Her research interests include optimization, linear algebra, scientific computing, and real-world applications. She is the coauthor of two books, *Practical Optimization* and *Numerical Linear Algebra and Optimization*, and the author or coauthor of many research papers. She has chaired the Advisory Committee for the Mathematical and Physical Sciences Directorate at the NSF, and the Advanced Scientific Computing Advisory Committee for the U.S. Department of Energy; she has also served on several other committees for the NSF and the NRC. She is a member of the scientific advisory board of the DFG Research Center "Matheon" (Berlin) and of the Center for Industrial and Applied Mathematics (Sweden). A member of both the NAS and NAE, she recently chaired the 2010 International Review of Mathematical Sciences Research in the United Kingdom. Dr. Wright is a fellow of the American Academy of Arts and Sciences; she also received a doctorate in mathematics (honoris causa) from the University of Waterloo (Canada) and an honorary doctorate of technology from the Royal Institute of Technology (KTH), Sweden.

Joe B. Wyatt served as chancellor and CEO of Vanderbilt University from 1982 to 2000. During that time, he led Vanderbilt's ascent into the top tier of U.S. teaching and research universities. He oversaw the expansion of the university's academic offerings, the diversification of the student body, and the increase of Vanderbilt's endowment from $170 million to more than $2 billion. Prior to joining Vanderbilt, Dr. Wyatt was a member of the faculty and administration at Harvard University, serving as vice president for administration from 1976 to 1982. During this period, he led EDUCOM, a consortium of 450 universities that developed computer networks and systems for sharing information and resources. In addition Dr. Wyatt co-authored the book *Financial Planning Models for Colleges and Universities* and wrote numerous papers and articles in the fields of technology, management, and education. Dr. Wyatt's earlier career focused on computer science and systems, beginning at General Dynamics Corporation in 1956, and continuing at Symbiotics International, Inc., a company he co-founded in 1965. Mr. Wyatt was a co-founder, vice chairman of the board, and chairman of the Investment Committee for the Massachusetts Technology Development Corporation, a public/private venture capital group that has financed many successful technology-based companies in Massachusetts. He is currently chairman of the board of the Universities Research Association.

Staff

Scott T. Weidman is the director of the National Research Council's Board on Mathematical Sciences and Their Applications (BMSA). He joined the NRC in 1989 with the Board on Mathematical Sciences and moved to the Board on Chemical Sciences and Technology in 1992. In 1996 he established a new board to conduct annual peer reviews of the Army Research Laboratory, which conducts a broad array of science, engineering, and human factors research and analysis, and he later directed a similar board that reviews the National Institute of Standards and Technology. Dr. Weidman has been full-time with the BMSA since mid-2004. During his NRC career, he has staffed studies on a wide variety of topics related to mathematical, chemical, and materials sciences, laboratory assessment, risk analysis, and science and technology policy. His current focus is on building up the NRC's capabilities and portfolio related to all areas of analysis and computational science. He holds bachelor degrees in mathematics and materials science from Northwestern University and M.S. and Ph.D. degrees in applied mathematics from the University of Virginia. Prior to joining the NRC, he had positions with General Electric, General Accident Insurance Company, Exxon Research and Engineering, and MRJ, Inc.

Michelle Schwalbe is a program officer with the BMSA and the Board on Energy and Environmental Systems (BEES) within the National Research Council. She has been with the National Academies since 2010, when she participated in the Christine Mirzayan Science and Technology Policy Graduate Fellowship Program with BMSA. She then joined the Report Review Committee of the National Academies before re-joining BMSA and later joining BEES. With BMSA, she has worked on assignments relating to verification, validation, and uncertainty quantification; the future of mathematical science libraries; the mathematical sciences in 2025; and the Committee on Applied and Theoretical Statistics. Her interests lie broadly in mathematics, statistics, and their many applications. She received a B.S. in applied mathematics specializing in computing from UCLA, an M.S. in applied mathematics from Northwestern University, and a Ph.D. in mechanical engineering from Northwestern University.

Thomas Arrison is a senior staff officer in the Policy and Global Affairs division of the National Academies. He joined the National Academies in 1990 and has directed a range of studies and other projects in areas such as international science and technology relations, innovation, information technology, higher education, and strengthening the U.S. research enterprise. He holds M.A.s in public policy and Asian studies from the University of Michigan.